BEAMS AND ACCELERATORS WITH MATLAB

Other books in this series by the author

One Hundred Physics Visualizations Using MATLAB (2013)

More Physics with MATLAB (2015)

Cosmology with MATLAB (2016)

Beams and Accelerators with MATLAB

The Companion Media Pack is available online at
http://www.worldscientific.com/worldscibooks/10.1142/10917#t=suppl

1. Go to /http://www.worldscientific.com/r/10917-suppl
2. Register /login/
3. You should be re-directed to / http://www.worldscientific.com/worldscibooks/10.1142/10917
4. Click on the Supplementary tab to download the Media Pack

BEAMS AND ACCELERATORS WITH MATLAB

With Companion Media Pack

Dan Green
Fermi National Accelerator Laboratory, USA

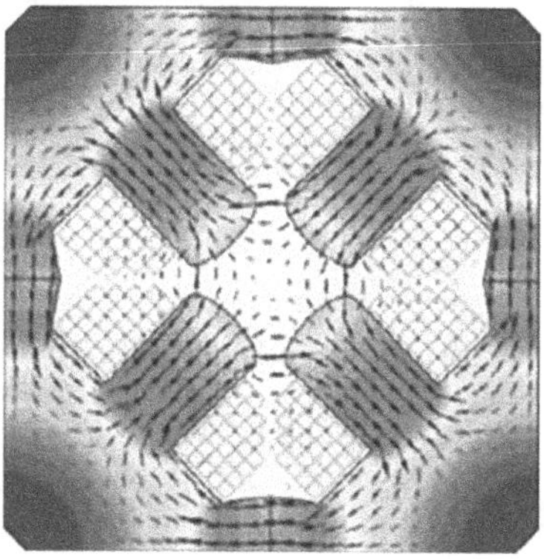

World Scientific

NEW JERSEY • LONDON • SINGAPORE • BEIJING • SHANGHAI • HONG KONG • TAIPEI • CHENNAI • TOKYO

Published by

World Scientific Publishing Co. Pte. Ltd.
5 Toh Tuck Link, Singapore 596224
USA office: 27 Warren Street, Suite 401-402, Hackensack, NJ 07601
UK office: 57 Shelton Street, Covent Garden, London WC2H 9HE

British Library Cataloguing-in-Publication Data
A catalogue record for this book is available from the British Library.

BEAMS AND ACCELERATORS WITH MATLAB
With Companion Media Pack

ISBN 978-981-3237-46-9

Preface

> If there is no solace in the fruits of our research, there is at least some consolation in the research itself. Men and women are not content to comfort themselves with tales of gods and giants, or to confine their thoughts to the daily affairs of life; they also build telescopes and satellites and accelerators and sit at their desks for endless hours working out the meaning of the data they gather.
>
> — **Steven Weinberg**

> I sometimes think about the tower at Pisa as the first particle accelerator, a (nearly) vertical linear accelerator that Galileo used in his studies.
>
> — **Leon M. Lederman**

The aim of this text is not to rigorously examine beam and accelerator physics. Many excellent texts exist for that purpose. Rather, the idea is to provide MATLAB scripts that embody "movies" of dynamical systems, "live" execution that can be followed directly from input to output, and "apps" where variables can be changed by the user with immediate graphical changes in output made available in real time. In this way the expectation is that the user of these scripts can build up an intuition in regard to beam and accelerator systems.

This text is not a formal exposition of the topics, since many authoritative texts exist (see references). Some heuristic derivations and expositions are presented here. However, the main objective is to enable the user of the scripts to interact with elements of beams and accelerators. To that end, interactive scripts are supplied on several topics and the user can supply input to see how the solution of a particular problem changes as the parameters of the topic change. The scripts may be direct MATLAB command line objects. However, in order to enhance the participation of the user, other types of script are commonly used. There are "apps" which define a GUI with tools to explore the real time evolution of systems. The graphics tools available in an app are presently rather limited. However, an alternative is a "live" script. Such scripts show the input commands in the editor window and the output in the command window simultaneously. In this way the user can more easily follow the course of the application script. In fact, any basic MATLAB script can be easily converted to and saved as a live script.

There are many numerical packages available with which to write analysis scripts. The author has chosen MATLAB because of the breadth of tools available, both for analysis and for graphical display. In addition, MATLAB is widely used in the engineering and physics departments of colleges and universities which allow, with sitewide licenses, free access to the MATLAB package. As an alternative, an economical student version of the package is available for purchase.

Knowledge of Maxwell's equations is assumed. In the text MKS units are used. However, for energies, the electron volt (eV) is used, to conform to beam and accelerator usage. In addition, equations will normally use the convention $c = 1$, where c is the speed of light. In that case, momentum (pc), mass (mc^2), and energy (ε) will all have the units of energy.

As regards relativity, the text assumes some prior knowledge. The relationship of mass (m), velocity (v), and energy (ε) is: $v = \beta c, pc = m\beta\gamma = \beta\varepsilon, \varepsilon = mc^2\gamma, \gamma = 1/\sqrt{1-\beta^2}$. In the future $c = 1$ will be used, since that is the conventional practice. The appropriate factors of c can be inserted for calculations by simply requiring the proper dimensions to be observed for any particular expression. As an aid to memory, m and p form the sides of a right triangle with hypotenuse ε. The symbolic math utilities available in MATLAB also make conversions from energy, momentum, and velocity simple, and examples are shown in the text.

Occasionally, reference will later be made to the Lorentz transformation. This operation applies to objects called "4-vectors", with four components. Examples are the transformation of space and time $(\vec{x}, ct)$ and energy and momentum $(c\vec{p}, \varepsilon)$. The transformation to an inertial frame moving with a relative velocity β is $x_4' = \gamma(x_4 - \beta x_3)$, $x_3' = \gamma(x_3 - \beta x_4)$, while components perpendicular to the relative motion are invariant. The velocity has a signed behavior depending on the direction of the transformation. The transformation along the x_3 axis takes a matrix form:

$$\begin{pmatrix} x_1' \\ x_2' \\ x_3' \\ x_4' \end{pmatrix} = \begin{pmatrix} 1 & 0 & 0 & 0 \\ 0 & 1 & 0 & 0 \\ 0 & 0 & \gamma & -\beta\gamma \\ 0 & 0 & -\beta\gamma & \gamma \end{pmatrix} \begin{pmatrix} x_1 \\ x_2 \\ x_3 \\ x_4 \end{pmatrix}.$$

For historical reasons the symbols β and γ also appear in conventional accelerator usage to represent quantities relevant to accelerator physics (physicists have limited symbolic imaginations). This duplication is made clear in the text and should not cause confusion.

Accelerators are associated with physicists in the popular imagination. However, the use of accelerators is much more widespread. Thus, there is in principle a large audience for tools which are

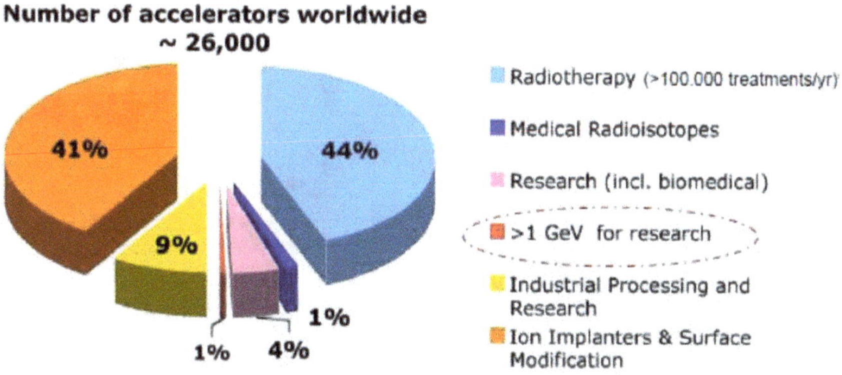

Pie chart of the applications of accelerators which are presently operating. Research applications account for only 1% of all accelerator operations.

designed to explain the operation of beams and accelerators. Indeed, there are many accelerators in operation which are not associated with physics. In fact, such applications are a distinct minority, as seen in the figure above.

The majority — 86% of all accelerators — are used either for medical applications or for industrial applications such as microelectronics.

Contents

Introduction to MATLAB

> The greatest value of a picture is when it forces us to notice what we never expected to see.
>
> — **John Tukey**

> There is a computer disease that anybody who works with computers knows about. It's a very serious disease and it interferes completely with the work. The trouble with computers is that you "play" with them!
>
> — **Richard P. Feynman**

MATLAB has an extensive suite of tools which are available to the user. The documentation is very good, so only a representative series of examples is given here. The user is strongly encouraged to first set up the program and then browse the documentation and demonstrations which are part of the MATLAB documentation, before setting out on any particular application.

Symbolic math and a matrix-based language make MATLAB a good vehicle for the study of beams and accelerators since both areas make extensive use of a matrix formulation of such systems. There are many MATLAB matrix utilities: matrix multiplication (*), determinant (det), inverse (inv), trace (trace), and solving the eigenvalue problem (eigen). These operations can be applied to both numeric and symbolic matrices, and will be used extensively in the scripts supplied with this text.

The text assumes a knowledge of vector operations. MATLAB provides many of those operations as utilities. Symbolic operations for dot product (dot) and cross product (cross) are supplied, as are numerical operations for "gradient", "divergence", and "curl".

The symbolic math package includes many trig identities, and a symbolic expression may be improved by using the symbolic operations "simplify", "simple", "factor", "expand", and "pretty". Sometimes the explicit operation "subs" can be used to simplify an expression when an identity eludes MATLAB.

Equations can be solved symbolically using "solve" for an algebraic system of equations and "dsolve" for a system of differential equations. Numerical solutions are available in the utility "ode45" for ordinary differential equations. For ode45 a simple script, "ode_demo", is provided in this text.

Once MATLAB is installed, a useful window layout is shown below. There are two main windows: the command window and the editor. There are also windows for the command line history and the variables in memory, which can be expanded as needed.

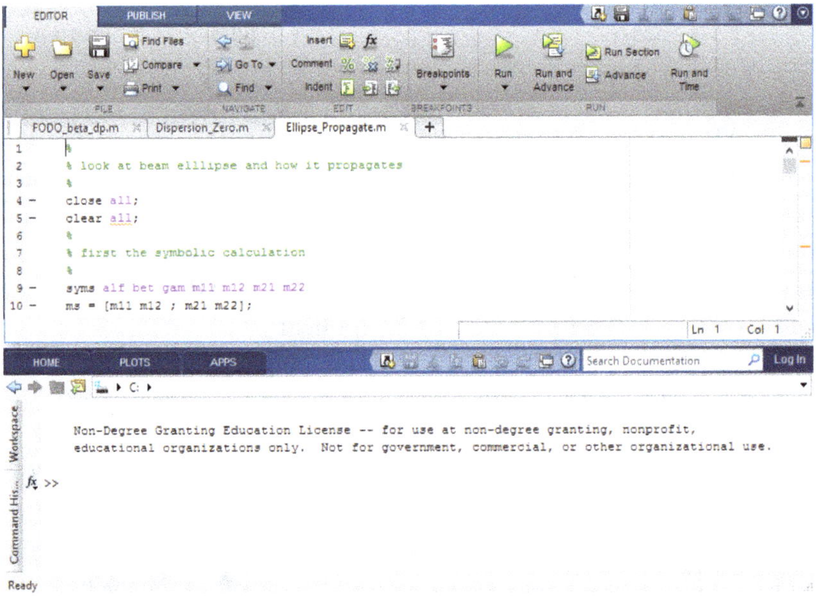

Figure I.1: Screen layout for MATLAB showing the main command and editor windows. The command history and the workspace variables are compressed on the left hand side.

In the command window there are tabs to open files, import data, and open plots. The command window Search tab is comprehensive. There is a command line help but the user needs to know the name of the utility to invoke it successfully. The command line prompt is $\gg$. The state of the computation is given by the "ready" or "busy" indication in the lower left corner of the command window.

The editor window is set up to create and edit scripts, "live" scripts, and "apps". It controls the running of scripts, with debugging breakpoint options. A sample of the editor display appears in the figure below.

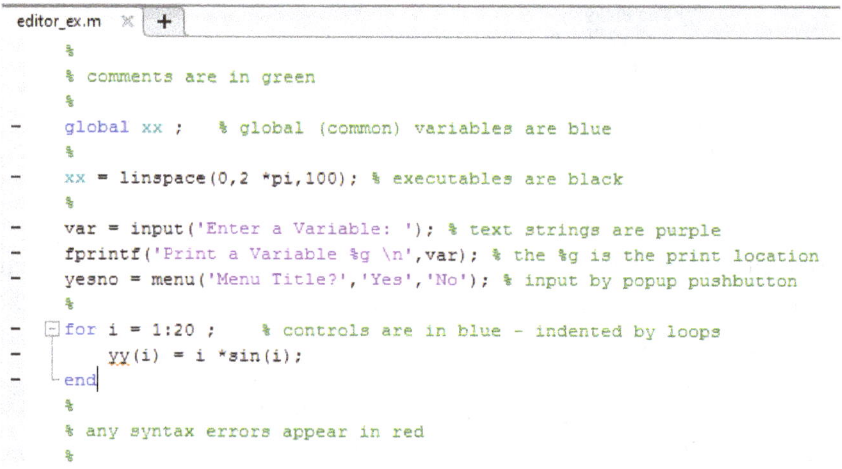

editor_ex.m
%
% comments are in green
%
global xx ;   % global (common) variables are blue
%
xx = linspace(0,2 *pi,100); % executables are black
%
var = input('Enter a Variable: '); % text strings are purple
fprintf('Print a Variable %g \n',var); % the %g is the print location
yesno = menu('Menu Title?','Yes','No'); % input by popup pushbutton
%
for i = 1:20 ;   % controls are in blue - indented by loops
    yy(i) = i *sin(i);
end
%
% any syntax errors appear in red
%

Figure I.2: Editor demonstration script showing the variety of commands and their color coding. Any errors of syntax typing of run time errors will appear in red. The script is provided in the text in "editor_ex". Execution time errors are shown in the command window.

The general categories of MATLAB functions can be invoked using the fx symbol in the command window. The result appears in Figure I.3. The tree structure can then be explored. The result of working down the tree for the "mathematics" category is given in Figure I.4. Essentially, all special functions are part of the MATLAB library.

A small set of simple scripts have been written in order to help a new MATLAB user to get started. The scripts are all simple

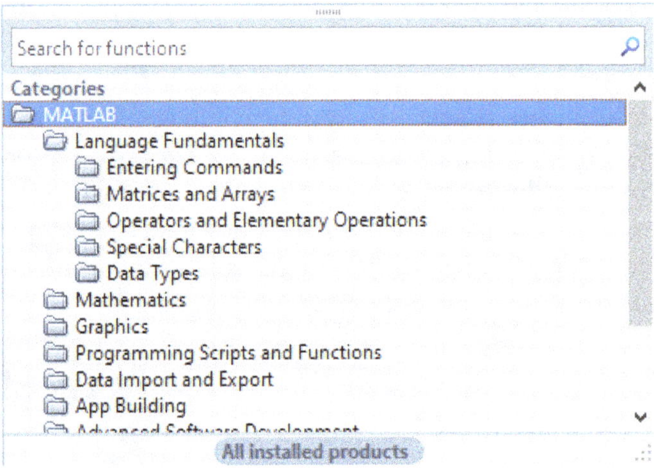

Figure I.3: General categories of MATLAB utilities. These categories can be expanded and explored.

".m files". The script "editor_ex" has already been shown. The script "plot_demo" makes a simple Cartesian two-dimensional plot as seen in Figure I.5. Other possibilities for plots include log, semilogx, semilogy errorbar, and loglog. Options for the symbols, colors, and line types used in "plot" appear in Appendix A.

The results of a computation are often displayed using the "plot" command. There are many commands associated with the basic "plot" command. Common ones are "figure" (which makes a new figure), "title", "xlabel", "ylabel", "legend", "axis", and "grid". Other types of two-dimensional plots include "bar", "hist", and "pie". Formats for these options can be found using the command line "help" facility. Multiple overlapping plots can easily be made using the "hold on" and "hold off" commands.

To edit a created plot, one can use the tabs in the figure taskbar. Choose file -> save, print and insert or tools -> edit plot, view plot editor. Formatting of the plot different from the defaults supplied by MATLAB is most easily accomplished in the plot editor.

There are also plots to display a function that depends on two variables. Relevant tools are "plot3", "mesh", "surf", and "contour". "Quiver" gives a visualization of the gradient. "zlabel"

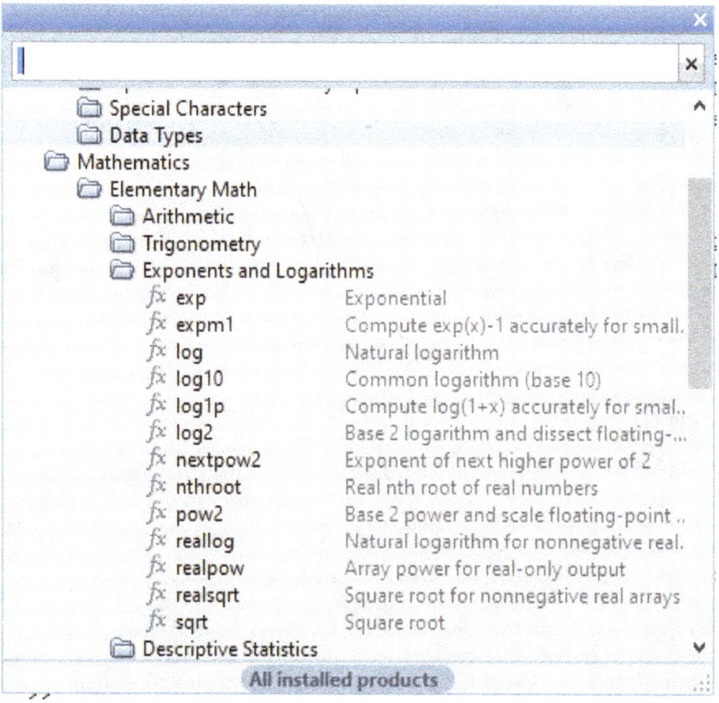

Figure I.4: Result of working down the tree structure of the mathematics category which appeared in the previous figure after expanding the exponents and logarithms subcategory.

and "meshgrid" are subsidiary tools. For reference purposes, the plot tab in the command window gives all the plot types, and the format for that specific type of plot also pops up.

The editing of a plot is similar to that for a two-dimensional plot. One particularly useful tool on the figure toolbar is a 3D rotation tool which lets the user change the point of view of the plot. There are many choices for colors. They can be displayed using the tab choice — insert -> colorbar. The edit -> colormap -> tools -> standard colormaps shows the already defined color maps. One fairly standard one is called "jet", which is used in this text.

More complex displays of results are shown in the supplied script "demo_3d". Some results of that script appear in Figure I.6.

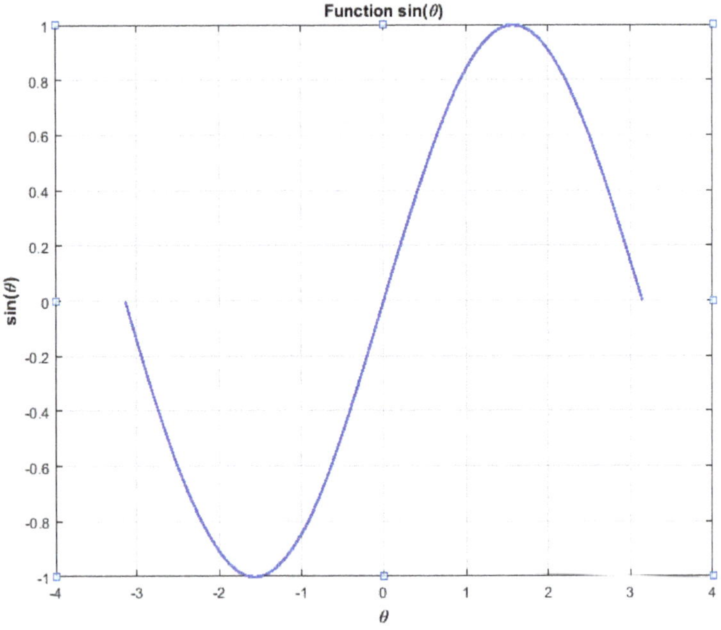

Figure I.5: Output plot created by the script "plot_demo".

There are flow control commands, shown in Figure I.7 for the demonstration script "template". This demo is supplied to illustrate program controls to start and stop a program under user control. The "while" and "break" commands enable the program control.

Some controls options are "while", "break", "if", "else", "elseif", "for", and "end". There are also logicals to assess decision points: equal (==), and (&&), or (||), less than (<), greater than (>), and not equal (~=). An example appears in the script "Template_2", some of whose commands appear in Figure I.7.

The editor enables the creation of files (.m files) to be written, compiled, run, and edited. The editor tabs are dropdown. The created .m files are saved. They can be executed in the Command Window or using the Run button in the editor. There are other options besides the .m file. There are GUI, Apps, and Live file types. This text will use mostly .m files, with some Apps and Live files as examples, depending on the applicability of each to the problem at hand.

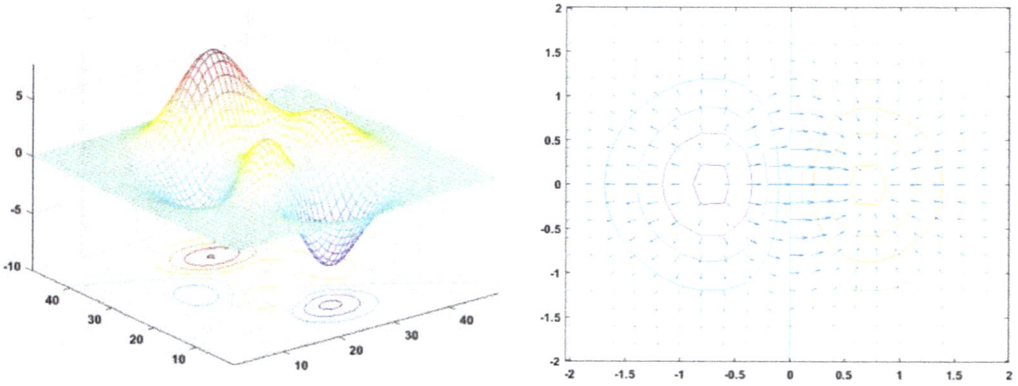


```
figure
[X,Y] = meshgrid(-8:.5:8);
R = sqrt(X.^2 + Y.^2) + eps;
Z = sin(R)./R;
mesh(Z);
%
figure
[X,Y] = meshgrid(-3:.125:3);
Z = peaks(X,Y);
meshc(Z);
%
figure
[X,Y,Z] = peaks(30);
surfc(X,Y,Z);
axis([-3 3 -3 3 -10 5]);
%
figure
[X,Y] = meshgrid(-2:.2:2);
Z = X.*exp(-X.^2 - Y.^2);
[DX,DY] = gradient(Z,.2,.2);
contour(X,Y,Z)
hold on
quiver(X,Y,DX,DY)
hold off
%
```

Figure I.6: Some output of the "demo_3d" script, showing the use of the MATLAB utilities "meshc", "gradient", and "quiver". The script commands appear on the right of the figure.

```
% Program to compute something
%
close all;              % clear all plots
clear all;              % Clear memory
help Template2;              % Print header
%
% put constants here
%
irun = 1;
%
while irun > 0
    kk = menu('Pick Another Option?','Yes','No');
    if kk == 2
        irun = -1;
        break
    end
    if kk == 1
        %
```

Figure I.7: Part of the editor display for the script "Template_2", which shows some aspects of the program control. The user input is enabled using the "menu" tool.

A "Live" field, ".mlx", is trivially created from a standard script, ".m", using the "save as" option in the "save" tab of the editor.

Numbers are defined in a format with an explicit decimal point or by "aeb", which denotes a x 10^b. Vectors are defined using x = [x1 x2 x3 x4 ....] as a row vector or x = [x1; x2; x3; x4; ...] for a column vector. Matrices are an obvious extension; M = [x1 x2 x3 x4; y1 y2 y3 y4] is a [2,4] matrix. A glance at the Workspace window will show what variables exist — symbolic or numerical and their dimensions.

Numbers can be added, subtracted, multiplied, and divided using the operations "+", "–", ".*", and "./". Matrices when multiplied use A*B. The transpose of a matrix A is A'. There are also special matrices: zeros, ones, eye (identity matrix).

Ordinary differential equations are solved numerically (Runge–Kutta) using "ode45". An example is provided in "ode_demo". For a numerical solution to an ordinary equation, the utility "fminsearch" is used in this text. Numerical equations where no analytic solution exists can be solved using this utility.

Some useful operations to perform on a vector are “mean”, “median”, “sum”, standard deviation (“std”), “max”, “min”, “prod”, “sort” , “length”, “diff”, and “gradient”. “Logspace” and “linspace” can be used to create regularly spaced vectors. “Trapz” supplies the numerical integral of a vector. “Quad” integrates, numerically, a user-defined algebraic function. “Int” integrates a symbolic function.

Symbolic variables are defined by declaring them in a “sym” command. They can be differentiated using “diff” or integrated using “int”. Their numerical values can be determined using the “eval” utility. The symbolic math package in MATLAB means that one need never again do a derivative, an integral, or solve a solvable differential equation. One can check the workspace to see that the variable is indeed symbolic. Symbolic variables are treated like text strings. Taylor expansion of symbolic functions is supported.

Expressions after symbolic computation are often ugly. Using the command “simplify”, “pretty”, “factor”, or “collect” can often make them look nicer. Numerical evaluation is done by supplying values of the symbolic variables and then invoking the utility “eval”. The resulting numerical values are useful in plotting symbolic expressions. The basic tool for solving symbolic differential equations is “dsolve”. The independent variable is fixed to be t. Constants are created as C if the initial conditions are not specified.

MATLAB provides basic polynomial fitting. However, there are no errors assigned to the data, which means that all points have the same statistical weight. The fits can also easily be done using the MATLAB utility “polyfit” instead of being accomplished dynamically using the figure toolbar with the tab “basic fitting”. Results are shown for an example in Figure I.8. The order of the polynomial can be chosen and the residuals can be plotted.

More than 60 scripts are supplied as an integral part of this text. They will be referenced as their specific topic is referred to in the text.

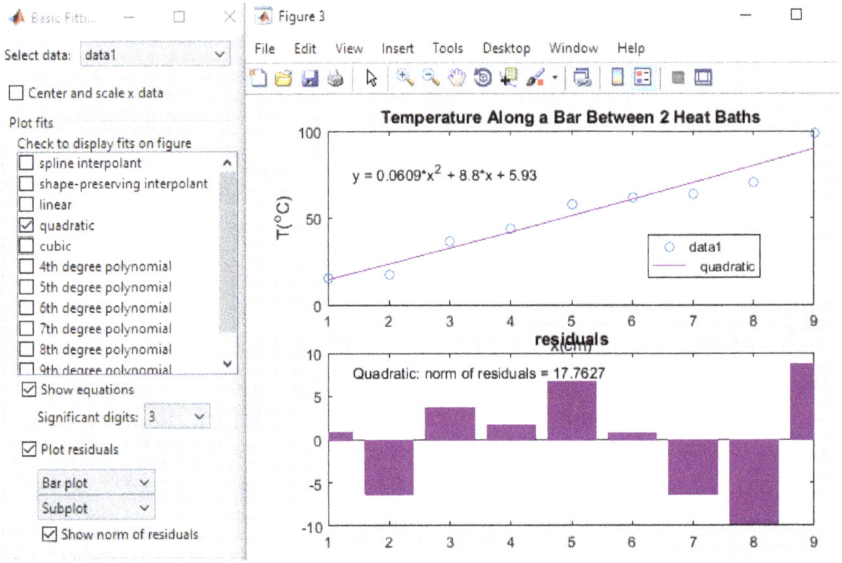

Figure I.8: Use of the figure toolbar in MATLAB in order to do simple spline or polynomial fits.

There are "movies" in the MATLAB documentation which are very useful. In particular, the movies on App and GUI creation in the MATLAB documentation are very valuable, and some information appears in Appendix B. There are also many other MATLAB "demos" that can be invoked. Often a search for a particular tool, e.g. "dsolve", leads to several useful examples.

"Movies" are often provided in the scripts. The independent variable is normally time, so the spacing of the frames on the movie indicates the velocity and acceleration of the system as it evolves dynamically.

Many of the symbolic tools of MATLAB use the "Maple" suite of utilities. Indeed, there are several accelerator tutorials which use the Maple tools exclusively. An example is the talk "Accelerator Physics Using Maple®", a course given at the US Particle Accelerator School (January 2016; Austin, Texas) by U. Wienands and E. Marin Lacoma (references). Because MATLAB has an expanded suite of utilities beyond what Maple provides, MATLAB has been the choice of analysis tools in this text. In particular, the belief that a picture, especially a dynamic one, is worth more than words informs this choice of software package.

This whirlwind tour has only scratched the surface of the available resources that MATLAB provides. The reader is encouraged to "play" with some of them in order to prepare for using — and modifying! — the scripts related to beams and accelerators provided in the text. A small set of the first scripts used in the text are appended in Appendix C for the reader to examine.

Chapter 1

Classical Optics

. . . the Whiteness of the Sun's Light is compounded all the Colours where with the several sorts of Rays whereof that Light consists. . .

— **Isaac Newton**

Most physicists have more experience with classical optics than with beam optics. For that reason it is useful to recall some facts and definitions from the former.

In classical optics the deviations from the axial line of a series of lens elements can be described by transverse position x and slope m. All transverse excursions are axially symmetric in classical optics, so that simple two-dimensional systems are possible. A matrix can be written that describes the action of some element of the optical system. A ray with an initial position and slope can be traced through the system by making successive matrix multiplications. The matrices for a free space ("drift") just represent a straight line given the absence of forces and a thin lens, concave or convex (focusing or defocusing), are

$$\begin{pmatrix} x \\ m \end{pmatrix} = \begin{pmatrix} 1 & L \\ 0 & 1 \end{pmatrix} \begin{pmatrix} x_o \\ m_o \end{pmatrix}, \quad \begin{pmatrix} x \\ m \end{pmatrix} = \begin{pmatrix} 1 & 0 \\ \mp 1/f & 1 \end{pmatrix} \begin{pmatrix} x_o \\ m_o \end{pmatrix}. \tag{1.1}$$

In the approximation of a thin lens, the position is not altered but the slope is increased or decreased by an amount x_o/f where f is the focal length. A distance f after the thin lens all incident parallel rays come to a point at $x = 0$. This is called parallel-to-point focusing. Reversing the process is called point-to-parallel focusing.

The Liouville theorem says that the volume of phase space, x^*x', where the slope m is hereafter replaced by $dx/ds = x'$ and s is the total path length of the central ray of the system, approximately distance z along the optic axis, is conserved in optical systems consisting of lenses and drift spaces. This requires that the matrix determinant be 1. The area in (x, x') space is constant, although it may change shape in passing through the optical system.

In point-to-point focusing, the M(1,2) element of the system must vanish. The M(1,1) element is the system magnification in this case. An example is the eye, with small magnification. In the case of parallel to point, M(1,1) = 0. An example is the telescope. For point to parallel, M(2,2) = 0. A familiar example is the flashlight. In classical optics the parallel-to-parallel condition requires M(2,1) = 0 with magnification given by M(2,2). An example is the Galilean refracting telescope. These matrix element conditions follow from the fact that the initial point has $x \sim 0$ while an initially parallel beam has $x' \sim 0$. These focal conditions will be seen again when particle beam optics is explored.

This treatment is very simplified. In classical optics there are aberrations due either to the basic lens shape or to variations in the shape because of a manufacturing error. Dispersion is also a common phenomenon, resulting from variation of the focal length with the wavelength of the incident light since the index of refraction depends on wavelength. In beam physics, by analogy, the effects of the electric and magnetic fields depend on the energy of the beam particle itself, which introduces dispersion in the orbits of "off-momentum" particles.

The variations in focal length are illustrated in the script "Spherical_Mirror". Incident parallel rays are traced through a spherical mirror and the "focal" point is displayed. For a thin lens it is seen that the focus is almost at a point, along the mirror axis at a distance half the radius from the mirror, but for thicker lenses the focal point is smeared out. The "menu" given to the user appears in Figure 1.1. By choosing the largest incident axial ray the user can explore when the spherical aberrations become large. The ray trace for a choice of

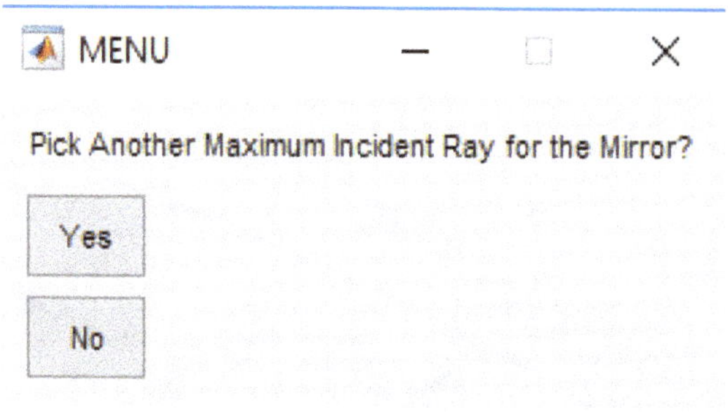

Figure 1.1: Menu for the script “Spherical_Mirror”.

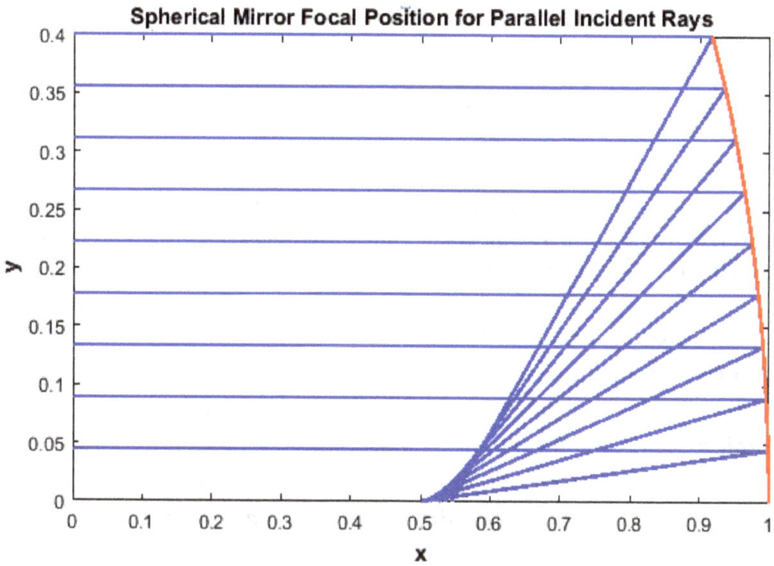

Figure 1.2: Ray trace for a spherical mirror illustrating the aberration inherent in a spherical mirror. The focal point smearing becomes worse for larger y value incident rays.

0.4 R appears in Figure 1.2. The focal point at x of 1/2 is already becoming smeared.

A spherical mirror is easy to grind but has this inherent problem. Astronomical mirrors with a diameter of about 10 m are now operational and larger ones are planned. They are often parabolic

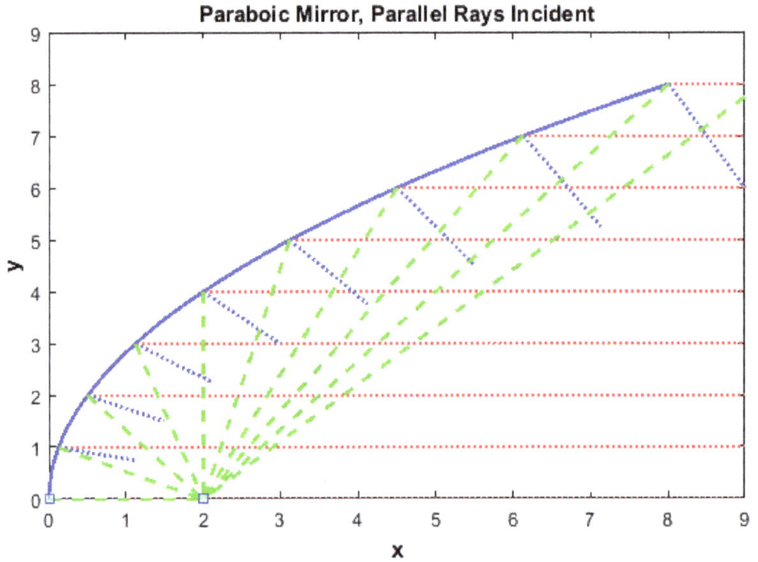

Figure 1.3: Ray trace for a parabolic mirror. The incident rays are red dotted lines, the normals to the parabola are blue dotted lines, and the reflected rays are green dashed lines. The focus is at $x = 2$.

in shape, because such a mirror, more difficult to manufacture, has no aberrations for incident parallel rays. There are still focal changes for off-axis rays, called "coma", but in astronomical applications the incoming starlight is quite nearly axial.

During the rest of the text, analogies with both classical optics and mechanics, especially simple harmonic motion, will become evident.

Chapter 2

Uniform Electric and Magnetic Fields

An external electric field, meeting it and passing through it, affects the negative as much as the positive quanta of the atom, and pushes the former to one side, and the latter in the other direction.

— **Johannes Stark**

The topics that will be explored are the motion of charged particles in electric and magnetic fields. These fields are the primary tools that are available to control beams of charged particles. They might make only a single pass through a system, defined to be a beam, or make many passes in stable orbits, as they do in accelerators.

The relativistic equation of motion is the Lorentz force equation, which states that the time rate change of momentum, p, is the force created by the electric (E), and magnetic (B), fields. The charge is expressed in units of the discrete electronic charge e, so that q is an integer. Vectors are indicated by the arrows over the symbols in question.

$$d\vec{p}/dt = qe(\vec{E} + \vec{\beta}c \; x \; \vec{B}). \tag{2.1}$$

The magnitude of the velocity is βc, the momentum p, the mass m, and the energy ε. They are connected, using $\gamma = 1/\sqrt{1-\beta^2}$, by the relationships

$$\vec{\beta} = \vec{p}/\varepsilon, \quad \varepsilon = \sqrt{p^2 + m^2} \tag{2.2}$$
$$\vec{p} = \gamma\vec{\beta}m = \vec{\beta}\varepsilon.$$

The formulae for these quantities will often be used treating c as being equal to 1, as in Eq. (2.2). This may seem disconcerting, but it is standard practice and energy units can be recovered using pc, ε, and mc^2 with $v = \beta c$. For example, $\varepsilon = \gamma m$ has the units restored by changing m to mc^2.

The equation for a uniform electric field — think of motion inside a large capacitor, oriented along the z axis — is solved symbolically in MATLAB in the live script "Uniform_E_Motion". The script uses the utility "int" to integrate the equation of motion and find z and "taylor" to expand the result in order to find the classical expressions. The utility "eval" is employed to make the numerical plots using the symbolic solutions. The solution for the velocity and position as a function of time is shown in Figure 2.1 and quoted in Eq. (2.3), where a is the classical acceleration qeE/m.

$$\begin{aligned} &dp/dt = qeE,\ p = (qeE)t \\ &\beta(t) = dz/dt = (p/m)/\sqrt{1 + (p/m)^2} = at/\sqrt{(at)^2 + 1} \\ &z(t) = [\sqrt{(at)^2 + 1} - 1]/a. \end{aligned} \tag{2.3}$$

```
Charged Particle in Constant Field Eo

dp/dt = q*Eo, p = (q*Eo)*t, beta = p/E = at/sqrt((at)^2 + 1), a = q*Eo/m



      2  2
sqrt(a  t  + 1)    1
--------------- - -
       a           a
Taylor Expansion for z

  3  4       2
 a  t     a t
- ----- + ----
   8       2
Classical Non-relativistic Results

bc = a t

zc =
```

$$\frac{a t^2}{2}$$

Figure 2.1: Output of the script showing the symbolic solution for $z(t)$ and the Taylor expansion applicable at short times of acceleration.

Since the time derivative of p is constant, p is proportional to t, and β is p/ε, the velocity at long times approaches c while the energy and momentum grow without limit. The classical results are recovered for short times, as is established using the MATLAB utility "taylor" to perform an expansion. The symbolic solution appears in Figure 2.1 and the position as a function of z is displayed graphically in Figure 2.2 for both classical motion and the correct relativistic motion. The script is a "live" one.

The results shown in Figure 2.2 for the velocity and position as a function of time appear dynamically in the script output as "movies". Each frame is at a fixed time interval, so that the observer of the movie can get a feeling for the actual velocity and position as a function of time. In the case of velocity, the classical value increases as t while the relativistic result asymptotically approaches c. In the case of position, the classical value increases as t^2 under constant acceleration while the relativistic result at long times increases linearly in t at a constant velocity of c.

For the case of a uniform magnetic field, a solution is first obtained symbolically. The direction cosines are defined to be dx/ds, where s is the arc length along the trajectory, $ds = vdt$. The magnitude of the velocity and the momentum remain constant in a magnetic field since the force is perpendicular to the velocity and does no work, so that

$$\begin{aligned} d\vec{p}/dt &= qe(\vec{v}x\vec{B}) \\ \vec{p} &= \gamma mv\vec{\alpha}, \quad \vec{\alpha} = d\vec{x}/ds \\ d\vec{\alpha}/ds &= qe(\vec{\alpha}x\vec{B})/p = qe(\vec{\alpha}x\vec{\alpha}_B/\rho), \ \ \rho = p/qeB. \end{aligned} \tag{2.4}$$

The direction cosines for the particle and the magnetic field are defined to be $\vec{\alpha}$.

For a uniform magnetic field oriented along the y axis, the momentum rotates in the (x, z) plane by an angle equal to s/ρ, where ρ is the radius of curvature of the circular orbit when the momentum is taken to be that perpendicular to the magnetic field. The positions describe a circle of radius ρ. The script "Uniform_By_Symbolic" is again a "live" (.mlx) file which allows the user to follow the computation line by line. It solves the Lorentz force equation

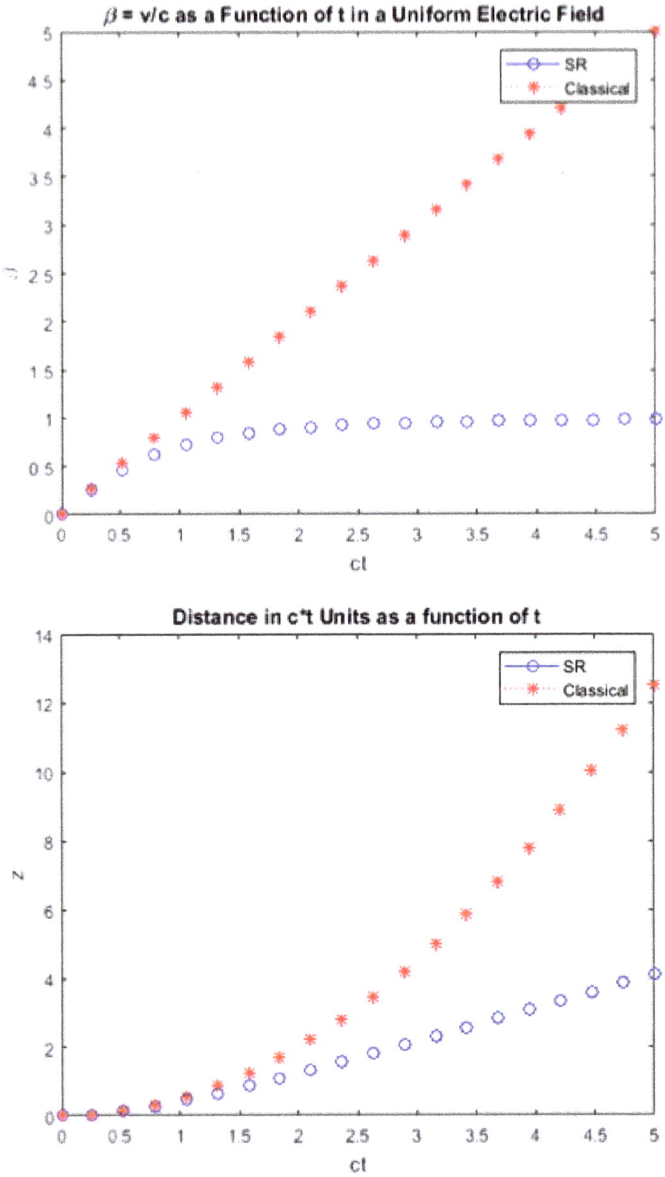

Figure 2.2: Solution for $\beta(t)$, $z(t)$ in the classical formulation and in the relativistic formulation, where β cannot exceed 1.

```
Charged Particle in Constant Field By

Initially moving along z axis, direction cosines

axs(t) =
```

$$-\frac{e^{-\frac{t\,i}{r}}\,i}{2} + \frac{e^{\frac{t\,i}{r}}\,i}{2}$$

```
azs(t) =
```

$$\frac{e^{-\frac{t\,i}{r}}}{2} + \frac{e^{\frac{t\,i}{r}}}{2}$$

```
Positions x and z
x(t) =
```

$$r\cos\left(\frac{t}{r}\right)$$

```
z(t) =
```

$$r\sin\left(\frac{t}{r}\right)$$

Figure 2.3: Output of the script "Uniform_By_Symbolic". The rotation in position by an angle s/ρ on a circle is evident. The cos and sin results for the direction cosines are not identified by the symbolic math package, which could be fixed by imposing a "subs" identity on the results.

symbolically, albeit with a convention which calls the independent variable s the time t, as is unfortunately fixed by the MATLAB utility "dsolve". The text output of the script appears in Figure 2.3. The direction cosines are found using the symbolic tool "dsolve" for differential equation solution, while the positions are obtained using the tool "int" and plots are made using "eval" for numerical evaluation.

For the exact numerical solution to having a charged particle in a uniform B field, the Cartesian solutions are provided in the script "Uniform_B_Motion". The functional input and output are shown in Figure 2.4, while the exact equations appear in Eq. (2.5), where the initial momentum vector is p_o and the initial positions are x_o, y_o,

```
function[x,p] = Uniform_B_Motion(s, xo, po, Bo)
%
% Take a step of arc length s in a uniform magnetic field of magnitude Bo
% along the y axis, charge qe, incident position,xo,yo, momentum components
% pox, poy poz - exact solutions

>> [x,p] = Uniform_B_Motion(1, [0 0 0],[0 0 10],1)

x =

    0.0015         0    1.0000


p =

    0.0300         0   10.0000
```

Figure 2.4: Initial parts of the script for a step in the field of arc length s with initial vector position and momentum x_o and p_o and output x and p, and an example where the units are "MKS" — meter (m), billion eV(GeV) and kilogauss (kG) with q the electronic charge.

and z_o. The rotation angle of the momentum vector in the $(x,\ z)$ plane is ϕ_B.

$$\begin{gathered}
\phi_B = s/\rho,\ \rho = p/qeB \\
\begin{pmatrix} p_x \\ p_z \end{pmatrix} = \begin{pmatrix} \cos\phi_B & \sin\phi_B \\ -\sin\phi_B & \cos\phi_B \end{pmatrix} \begin{pmatrix} p_{ox} \\ p_{oz} \end{pmatrix} \\
y = y_o + s p_{oy}/p_o \\
\begin{pmatrix} x \\ z \end{pmatrix} = \begin{pmatrix} x_o \\ z_o \end{pmatrix} + \rho/p_o \begin{pmatrix} p_z - p_{oz} \\ p_x - p_{ox} \end{pmatrix}.
\end{gathered} \tag{2.5}$$

The units used in the text are normally m, GeV, and kG. In MKS units the radius of curvature of a particle with q units of e, when p is in GeV, B in kG, and ρ in m, is

$$\rho = p/qeB, \quad e = 0.030. \tag{2.6}$$

For example, a particle with 100 GeV momentum in a one tesla or 10 kG magnetic field has a radius of curvature of 335 m. Since the velocity is almost c, the period of rotation is 7 μsec with a circular frequency of 893 kHz. In this text q is implicitly equal to 1 since only electrons and protons are typically considered. However, q can be any positive or negative integer in general.

It might be asked: Why is there a focus on magnetic fields in the literature? The simple reason is that magnetic fields which make substantial changes to the trajectories of relativistic particles are easier to create than the equivalent electric fields. The ratio of the bending power is approximately $300\beta B[T]/E[MV/m]$. For an electric field of 1 MV/m and a B field of 8 T, the magnetic force for $\beta = 1$ is 24 times more effective.

If B is taken in T (MKS units for the field, weber/m^2 or V*sec/m^2) and p is taken in TeV, then $e = 0.00030$. For the Fermilab Tevatron, the radius is about 1 km. With conventional magnets of 1.6 T the ring maximum energy is about 0.48 TeV. With 4 T superconducting magnets about 1.2 TeV could be obtained. The CERN LHC has a radius of about 4.3 km and improved magnets operating at about 8 T. The maximum energy of the LHC can then be estimated to be about 10 TeV. It is presently 7 TeV because the machine is not full of only dipole magnets. In what follows, R is used as the radius of an accelerator and is taken to be approximately ρ.

For any arbitrary field one can take small steps in arc length rotating to the frame where the field is along y and then rotating back to the original frame after taking the small step. In this way, transporting through any arbitrary field is numerically enabled. In practice this procedure is quite computer-intensive and it is perhaps easier to use a numerical solution such as Runge–Kutta integration over a grid in space with a detailed magnetic field map.

The full three-dimensional helical orbit is shown as a specific movie by running the script "Uniform_B_Motion_2". The last frame of the movie appears in Figure 2.5.

Very approximately, the helical orbit in a dipole can be characterized by a bend by angle ϕ_B at the centerline of the dipole and a straight line in (y, s) space rather than a helix where L is the length of the magnet in Eq. (2.7). This approximation is useful

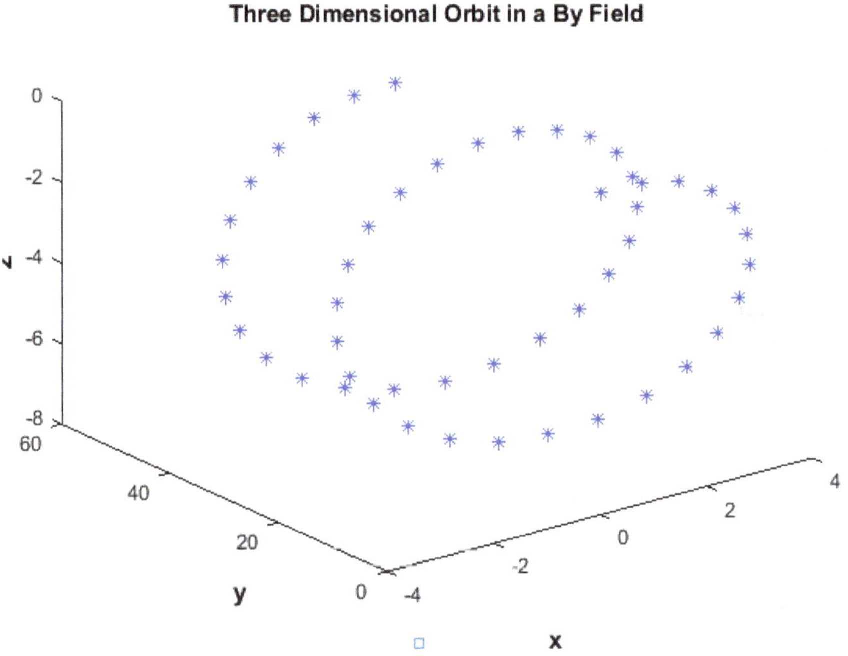

Figure 2.5: Helical path in three dimensions for a particle orbit in a magnetic field oriented along the y axis.

as an initial estimate and also because the orbits that appear in high-energy-physics applications are often due to "thin dipoles".

$$
\begin{aligned}
\phi_B &= L/\rho \\
y &\sim y_o + y_o' L \\
x &\sim x_o + x_o' L + \phi_B L/2.
\end{aligned}
\tag{2.7}
$$

The bend angle imparted by the dipole can be considered to be a transverse momentum impulse, $\Delta p_{TB} = eBL$, $\phi_B = \Delta p_{TB}/p$, imparted to the particle which is proportional to the product BL. The product BL can be thought of as the overall "strength" of the dipole. These approximations can be evaluated in the particular case displayed in Figure 2.4 for the exact solution. For a "thin" dipole they are a reasonable approximation.

The main use of dipoles is to steer the beam into a desired path. Other uses are in conjunction with the use of collimation to momentum-select the particles in a beam, since the bend angle depends on the particle momentum. Unwanted particles can be swept away and absorbed. The physical dispersion in x of a beam with momentum can also be actively employed to measure the momentum of a beam particle using detector elements placed in the beam itself.

The combination of a uniform electric and magnetic field is used in beamlines to select the mass of a particle in a mixed beam where the momentum is defined by the use of dipoles and collimators. This application may be considered to be a variant on the well-known experiment of J.J. Thomson, who measured the charge-to-mass ratio of the electron using electric and magnetic fields simultaneously.

In the case of a vertical electric field, E_o, and a magnetic field along the x axis, B_o, with a beam incident along the z axis, the net force is along the vertical y axis: $F_{ExB} = qe(E_o - \beta_o c B_o)$. There is no force acting on those beam particles — a mixture of pions and kaons, for example — which have the velocity $\beta_o = E_o/cB_o$. For a region of length L where both fields exist, the vertical force causes an angular displacement:

$$\begin{aligned} F_y &= dp_y/dt, \; dt \sim L/\beta c \\ d\theta_y &\sim dp_y/p_o \sim qeE_oL/cp_o(1/\beta - 1/\beta_o). \end{aligned} \tag{2.8}$$

For a mixed beam of charged pions, mass 0.1396 GeV, and kaons, mass 0.4937 GeV, a typical beam is that operated at the JPARC facility in Japan. The central momentum is 1.1 GeV. Pions then have a β of 0.992 while kaons have a value of 0.912. With a voltage potential of 550 kV in a 10 cm gap, E_o is 5.5 MV/m; with $L = 0.4$ m and a magnetic field of 0.02 T, the value for β_o is 0.916 and kaons are undeflected. The pions are deflected by 0.18 mrad. If they are then focused on a collimator slit at a downstream distance of 15 m, the pions and kaons are separated by 2.7 mm. A plot of the distribution in y of pions and kaons at this “mass slit” in this JPARC electrostatically separated beam is shown in Figure 2.6.

The beam has a finite spread of momentum, so chromatic aberrations created by the dipole magnets must be well canceled in order to achieve a focus spot size smaller than the separation. For

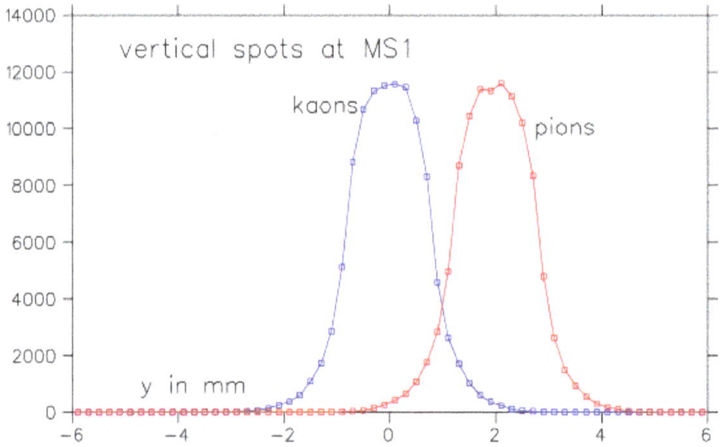

Figure 2.6: Vertical beam spot sizes in the JPARC kaon beam. Note that, in reality, the pions are much more copious than the kaons, so that a pure kaon beam is not truly achieved, but only enriched. The separation of the centroids is consistent with the value estimated above.

this reason sextupole lenses are used to make those corrections. They will be mentioned later, in the discussion of particle beams.

Successful use of this technique is limited to low energy beams. The cause of that limitation is the fact that all particles, regardless of mass, approach a limiting velocity of c at high energies, as seen in Figure 2.2a. Defining the deviation of the velocity from this limiting value to be $\delta\beta$ or $\beta = 1 - \delta\beta$, the change in the velocity depends inversely on the beam momentum $\delta\beta = 1/2(m/p_o)^2$, so that the physical separation degrades rapidly at high momentum.

Similar difficulties afflict other particle identification methods, such as Cerenkov radiation and time of flight (TOF), as will be mentioned in the section on particle beams. These techniques, which are used to tag the mass of a particle, pion, kaon, or proton, if they rely on a measurement of velocity, become degraded at high energies. In fact, for the highest energy experiments, particle identification by these means is not often attempted.

Chapter 3

Dipole

Magnetism, as your recall from physics class, is a powerful force that causes certain items to be attracted to refrigerators.

— **Dave Barry**

Previously, the approximation of a perfectly homogeneous magnetic field covering all space was assumed. It is clear that a real field is rather more complicated. A simple method of creating a field is to employ a current loop. The numerical constant needed is μ_o, which is 1.26×10^{-6} T*m/Amp. The field of a current loop evaluated along the axis, z, of the loop has a closed form solution:

$$B = \mu_o I R^2/[2(R^2 + z^2)^{3/2}], \tag{3.1}$$

where I is the loop current, z the distance along the loop axis, and R the loop radius. For example, at $z = 0$ with $I = 10{,}000$ A and $R = 5$ cm, the field is 0.125 T.

As a further look at several current loops, a current loop is solved for numerically using the Biot–Savert law by integrating over the loop current elements. There are options given in the script "Solenoid_4"; the z spacing between loops and the number of loops (one, two or four) as seen in Figure 3.1. The matrices for the field calculations use the utility "zeros" to set all elements initially equal to zero.

The dialogue with menus used in the script appears in Figure 3.2.

The MATLAB tool "menu" was employed to allow the user to choose parameters without using command lines to specify the input. The script "Solenoid_4" uses the plotting tools "contour" and "mesh", which were mentioned in the introduction.

```
>> Solenoid_4
  Program to plot B field of a single loop, a pair of current loops, or 4 loops
  numerically - all points using Biot-Savert


 B Field for 1, 2 or 4 Current Loops Pair, Radius 1, Separation = 2d
 Current Loops in x,y Plane. Theta is the Angle w.r.t. the z Axis
 B Field for a Current Loop, Radius a, for r >> a, Biot-Savert
 One Loop Dipole Moment ~ pi*I*a^2
Enter the Distance Between Current Loops ( ~ 0.5): 0.5
```

Figure 3.1: Numerical evaluation for the magnetic field of a current loop. There can be one, two or four loops and the spacing between them is an input parameter.

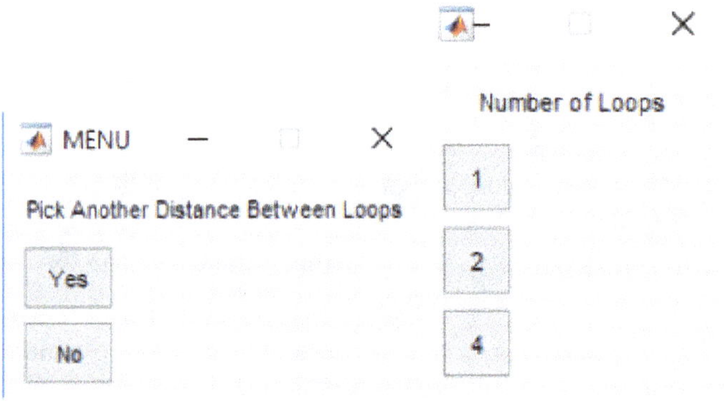

Figure 3.2: The dialogue used in "Solenoid_4" to start and end the script and to choose the number of current loops to plot.

The result for a set of four current loops, spaced apart by half the loop radius, appears in Figure 3.3. It is seen that there is a quite uniform field along the z axis, which then falls off as one goes outside the loop in z. In addition, the field is largest near the current sources. The x component of the field is also computed in the script and a contour plot is provided to the user and plotted in Figure 3.4. That field component is small near the center of the array of four loops, but is large at the current sources and outside the volume of the loops.

The different magnetic elements can all be thought of as superpositions of the elements of a set of individual current loops. There are two distinct forms for the current loop in actual practice. If they are added together as in the script, a solenoid is created. The field of

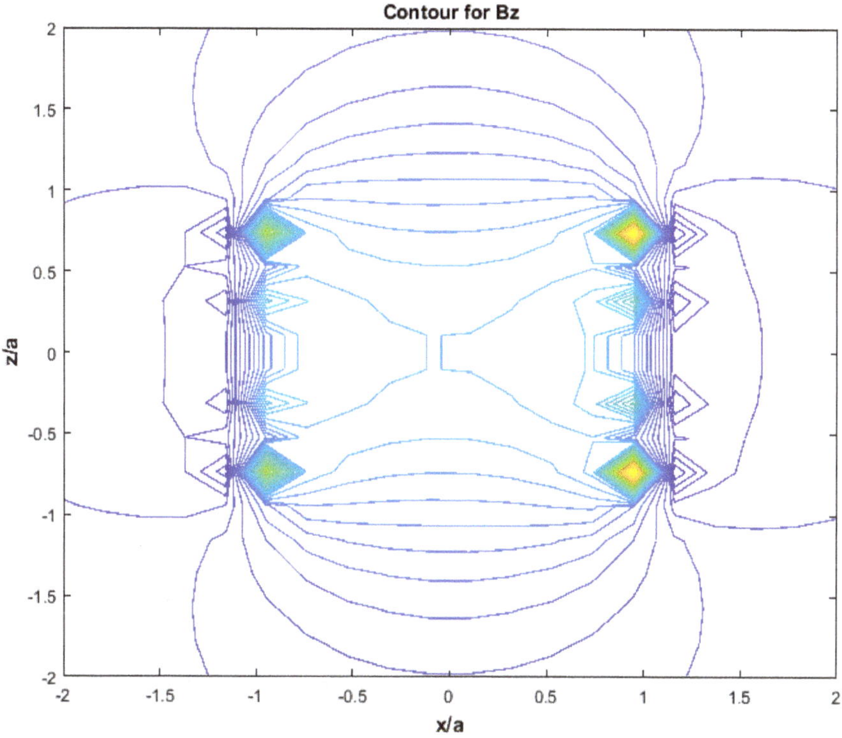

Figure 3.3: Contour plot for B_z with four current loops spaced by half the radius of the loops.

a very long solenoid, with a tight set of loops to reduce flux leakage, is $B = \mu_o \ln$, where n is the number of coil turns per unit length. For example, the CMS solenoid at the LHC has a coil of four layers 2 cm thick, each carrying 16 kA, resulting in a field of 4 T.

If the circular loop of a simple Helmholtz coil is stretched along the beam axis, a dipole field is created which approximates a uniform field in the vertical direction — the configuration assumed previously. Soft and permeable iron is used to further shape the resulting field in what are called "conventional magnets". In particular, the iron captures the magnetic flux lines and drastically reduces the "fringe field" which exists outside the volume defined by the region between the two coils as long as the iron is not saturated, or for fields less than about 2 T.

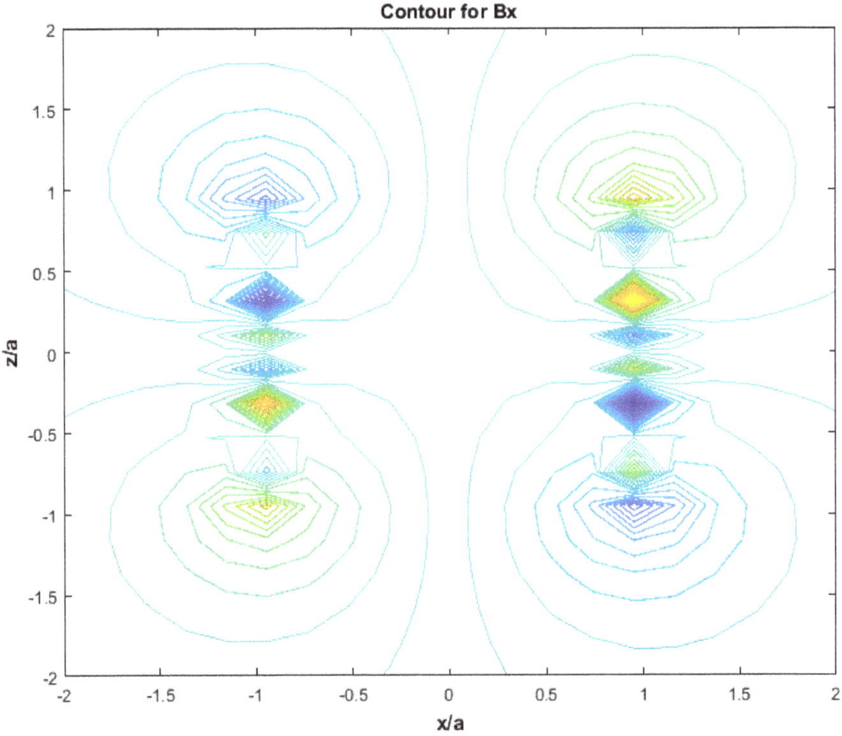

Figure 3.4: Contour plot for B_x with four current loops.

More recently, magnets made of superconducting elements have been constructed for two basic reasons. Higher fields can be obtained and the power dissipation in a superconductor is zero, so the power bill is much reduced. Fermilab created the first accelerator using superconducting elements. The initial Main Ring used conventional magnets and could obtain a proton beam of 400 GeV. The Tevatron used superconducting magnets in the same tunnel with the same radius and routinely operated at 900 GeV.

The existence of several field components is an inescapable feature of Maxwell's equations. This means that there are always "fringe fields" since the field cannot vanish immediately. Imagine creating a "dipole" by elongating two current loops along one axis, such as the z axis. The current loops are separated in x and y, and the currents in the two loops reinforce the field in the y direction. There will then be a field along the y axis which is reasonably constant inside the region between the loops and extends over some distance along z.

Maxwell's equations for the magnetic field in the absence of sources and time-varying fields are $\vec{\nabla} \cdot \vec{B} = 0$, $\vec{\nabla} \times \vec{B} = 0$. Soft iron can be deployed outside the current loops to capture and reduce the leakage fields or fringe fields. For example, if the rundown from B_o of the main field B_y from a dipole is linear in z over a distance d, then

$$\vec{\nabla} x\vec{B} = 0, \quad \partial B_y/\partial_z = \partial B_z/\partial_y = B_o/d \Rightarrow B_z = B_o y/d. \qquad (3.2)$$

The z component of the field is negative for negative z and positive for positive z. The x component of the field follows from the vanishing divergence and requires a nonzero field which is approximately quadratic in the distances transverse to the beam direction: $\vec{\nabla} \cdot \vec{B} = 0$, $B_x \sim -B_o xy/d^2$.

A longitudinal field is induced with a strength dependent on the y deviation from the reference trajectory. It can act as an entrance and exit lens to the dipole. There is no field on the y axis. The "effective" length of the magnet, as regards the overall momentum impulse, is approximately $L + d$.

A magnet of this type is called a dipole. The two "poles" or current loops reinforce in the vertical direction and extend over a distance L along the z direction. The length, physical or magnetic, of magnetic elements will be designated as L in general, while the lengths of field-free regions, or "drifts", will normally be defined to be d in what follows. A schematic of such a dipole appears in Figure 3.5.

The dipole shown in Figure 3.5 is a "conventional" magnet. Superconducting magnets use less iron and more current-carrying superconducting wire.

In the Tevatron, protons and antiprotons are stored and collided. Since the Lorentz force for magnetism is invariant under a sign change of both q and β, the two can exist within the same beam pipe and have the same but counterrotating orbits.

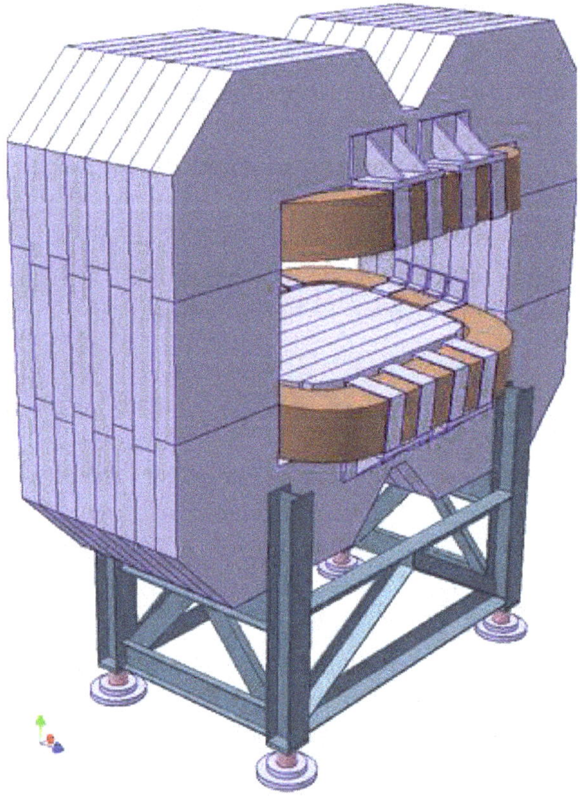

Figure 3.5: A conventional dipole magnet which uses current loops and iron to shape the field.

In the case of the LHC, protons collide with protons, necessitating two distinct orbits and magnetic fields. In this case superconducting dipoles are used in a "2 in 1" configuration, as shown in Figure 3.6. The magnet contains two distinct vacuum pipes where the magnetic fields are vertically oriented, but one is up and the other is down. Not shown in the figure are the restraints on the magnet. These are needed because the coils violently repel one another and need to be restrained in order to create and preserve the quality of the dipole fields. The stored energy in the field is $U = B^2/2\mu_o$. For the LHC dipole it is about 7.8 MJ, or the energy of a 20 T truck moving at about 60 mph. Obviously, since these magnets are superconducting,

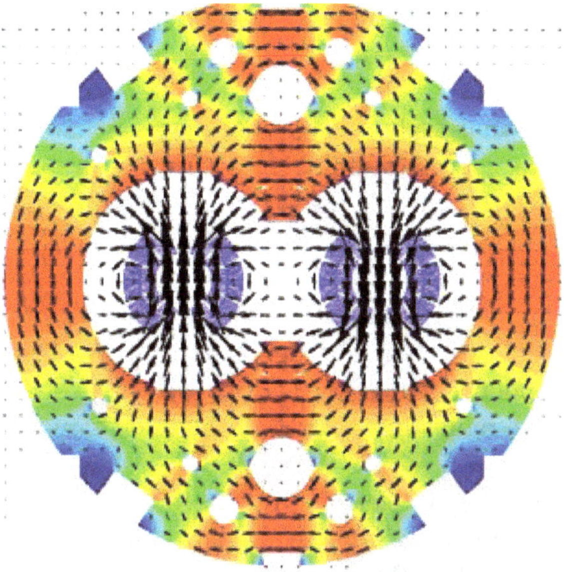

Figure 3.6: Map of the fields of an LHC dipole, showing the "2 in 1" solution for the two proton beams. The field is 8.6 T, which is a factor of approximately 2 greater than the Tevatron dipoles.

care must be taken to insure that they do not become normal conductors and discharge all that energy.

Simply spinning in a circle is not that interesting. The energy needs to change; one hopes that it can be increased. An early use of a uniform B field and a radio frequency (RF) source of electric field for particle acceleration was the cyclotron. Particles are supplied by a source and are emitted almost at rest. They are bent in a circle until they reach the "dee", where they are accelerated. When they reach the next dee, the electric RF field must change sign in order to continue to accelerate the beam. Therefore, an RF oscillation of the field is needed.

That is simulated using the utility MATLAB "round" to identify the dee. In the case of nonrelativistic motion, the circular frequency is $\omega = qeB/m$, where q is an integer and the radius of the orbit is $\rho = p/qeB$, where p is the momentum. The important point is that the frequency of the applied power is independent of the momentum of the particle. For a voltage in the RF of V_o synchronized to be the same during particle passage of the dee gap for each traversal, the particle gains a total kinetic energy of $2V_on$ in n complete turns or rotations.

The simple analysis breaks down as the particle energy increases, because the frequency scales inversely with the total energy divided by the mass of the particle: $\omega \to \omega/\gamma$, $\gamma = \varepsilon/m$. The solution to this problem in synchrotrons will be discussed later, in the section on acceleration.

The script "Cyclotron_NR" shows a "movie" of the particle trajectory in a cyclotron. For nonrelativistic motion the rotation frequency is independent of the particle velocity, so the RF can be a constant. The radius increases with the velocity, until the particle exits the cyclotron.

The dialogue for "Cyclotron_NR" appears in Figure 3.7, while the orbit appears in Figure 3.8. Note that the points from the movie have a fixed time interval, so that the velocity increase during acceleration shows as an increasing distance between points after an accelerating gap between the dees is traversed. This illustrates the utility of the movies in establishing an intuition about particle orbits. To set a scale, for a proton in the cyclotron, the circular frequency is 95.5 MHz times B in T. Hence the convention of calling the electric field RF, since this is in the FM band.

A schematic of a cyclotron appears in Figure 3.9. The pole tips of the dipole electromagnet supply the uniform magnetic field. The dees are conducting vacuum vessels which supply the electric field in the gaps. The particles exit the vacuum chamber at an energy defined by the magnetic field and are available to strike a target for experimentation. The electric field is oscillatory, so that the beam

```
Cyclotron w = qB/m, r = m*v/qB = v/w, E = (qBr)^2/2m
Relativistic Effects, m -> E, w Decreases and r Increases by gamma
Enter Number of Half Rotations, ~ 40: 40
Enter Energy Kick in Cross Dees ( ~0.5, dE~rdr): 0.5
Exit Dees
```

Figure 3.7: Output from the dialogue in "Cyclotron_NR". The "Exit Dees" message means that the proton has exited the cyclotron structure within the selected number of half turns and the energy increment per "dee" traversal.

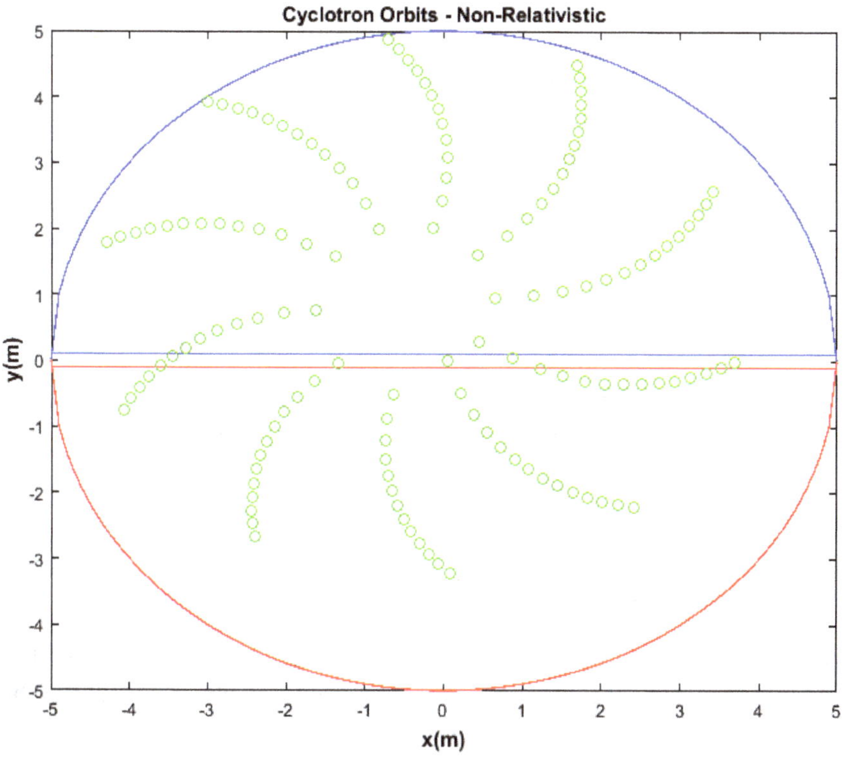

Figure 3.8: End of the movie for a cyclotron, when the beam exits the "dee". The proton is injected in the center of the cyclotron and gains energy after each traversal of a dee.

is "bunched" in phase with the accelerating portion of the voltage. This property of bunched beams will be explored in detail later.

As a numerical example, for a field radius of 0.53 m and a field of 1.6 T, the rotation frequency for deuterons is 12 MHz and the final kinetic energy gained is 17 MeV. The limitation for the cyclotron is simply the maximum magnetic field that can be supplied over the largest radius as long as relativistic effects remain negligible.

The way to avoid the issue of limitations is to confine the beam to a small aperture, apply the magnetic field to that aperture, and expand the radius using the ring of magnets so defined. This scheme means that for high energies the magnetic field must increase (ramp) as the particle energy increases and the RF must change

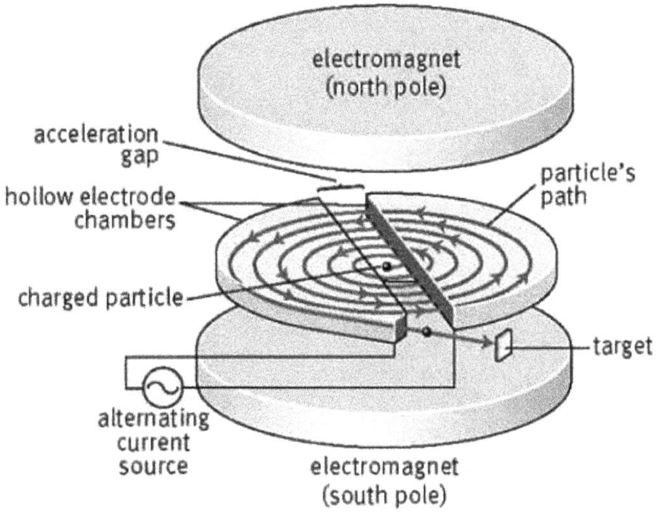

Figure 3.9: Schematic view of a cyclotron magnet, acceleration gap, RF, dee, and extraction of the beam to strike an experimental target.

to accommodate the velocity change due to the acceleration of the beam. The orbit in the ring will then remain constant while the energy increases. These more complex aspects will be explored in the section on acceleration. For a constant ρ as p increases, B must increase in lockstep in order to keep the beam within the vacuum aperture.

Once the maximum B field is achieved, the beams may be stored, with the RF field turned off to make the ring a "storage ring". This idea is exploited in colliders such as the LHC at CERN and the Tevatron at Fermilab. A second possibility is to extract the maximum energy beam and supply it (called the primary beam) directly to experiments or to strike an external target to produce secondary particles of interest to the experimenter, such as pions or kaons. At Fermilab this extracted beam was supplied at 400 GeV by the Main Ring or at 150 GeV by the more modern Main Injector.

Chapter 4

Quadrupole

Reality is what kicks back when you kick it. This is just what physicists do with their particle accelerators.

— **Victor J. Stenger**

Arrays of charges and currents can possess higher order multipoles beyond the monopole and dipole configurations which have been examined so far. An electric quadrupole can be devised with four charges arrayed as shown in Figure 4.1. There is no field at the origin. By symmetry there is no y field along the x axis and no x field along the y axis. The charges are positive (o) and negative (*).

There is a restoring force on a particle with an excursion in x, but an antirestoring force along an excursion in y. This means that the "lens" is "focusing" in x but "defocusing" in y. Such a system is not used in practice, but rather a magnetic quadrupole is constructed using four current loops appropriately arranged and with currents appropriately directed. This object, called the quadrupole, is the basic lens element used in beams and accelerators.

The quadrupole plays the role of the lens in optics much as the dipole plays the role of the optical prism. The main difference is that simultaneous focusing is not possible with a single quadrupole because there are no isolated magnetic charges such as exist with isolated electric charges.

Because there are no charges or current in the volume which contains the beam, fields can be considered which are derivable from a scaler magnetic potential, Φ. The beam charge and current are weak enough to be ignored, for now. The potential and the magnetic

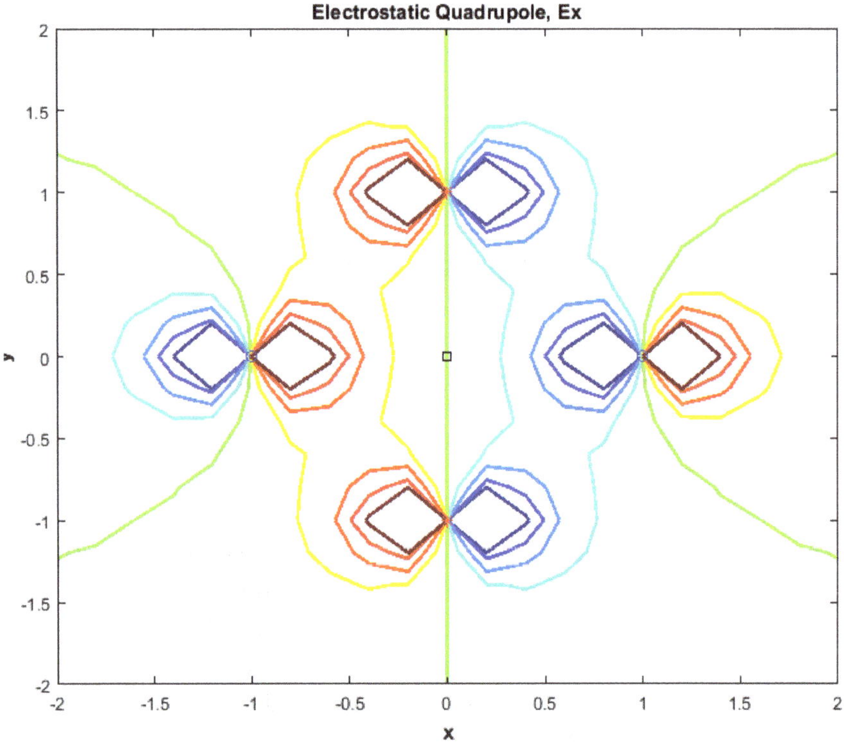

Figure 4.1: Contour of the E_x electric field for an electrostatic quadrupole. The centerline is indicated by the black square; the charges are blue o (positive) and red * (negative). The E_y field is the same as E_x after a rotation of the axes.

field in the case of a magnetic quadrupole are

$$\begin{aligned}\Phi_Q &= (dB/dr)xy \sim \cos(2\phi)r^2 \\ \vec{B}_Q &= \vec{\nabla}\Phi_Q.\end{aligned} \tag{4.1}$$

The quadrupole is characterized by the radial derivative of the field, dB/dr, called the gradient, which is arranged to be zero on the magnetic centerline. The x and y coordinates are perpendicular to the direction of the reference particle orbit, here taken to be the z axis. It is convenient and conventional to use the more precise arc length on a reference orbit, s, instead of z or time, t, as the independent variable. This was already shown to be simplifying for the case of the

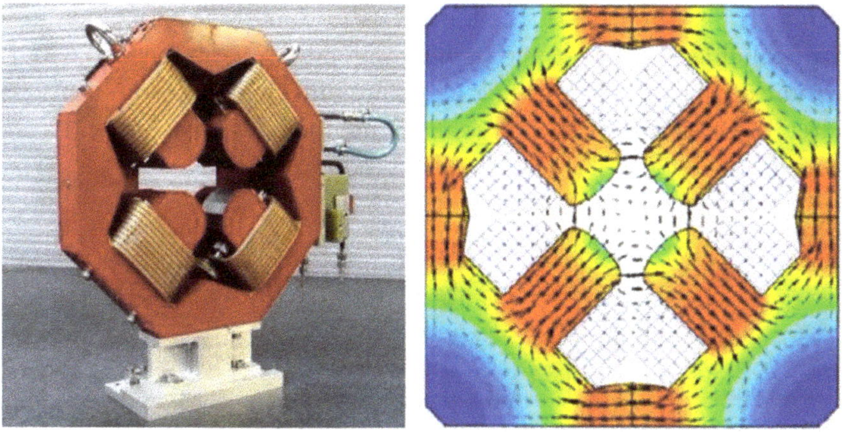

Figure 4.2: A quadrupole lens showing the current-carrying coils and the iron used to shape the field. A vacuum beam pipe would normally be aligned with the geometric center of the quadrupole. The field lines showing the effects of iron shaping appear on the right.

dipole, where the use of s made the dy/ds trajectory a straight line, rather than a helix if z is used.

A photo of a quadrupole is shown in Figure 4.2 with the current-carrying coils, the field-shaping iron, and current and cooling water leads. Using high quality iron, the pole tip is approximately an equipotential, which is why the poles are hyperbolic in shape, following the contour of Eq. (4.1). The magnetic field is then perpendicular to the surface of the iron.

It is assumed that the reference orbit, differential length ds, is defined by the dipole magnets and lies in the (x, z) plane, with y perpendicular to that plane. For small deviations from the reference orbit, the motion is characterized by x, $dx/ds = x'$, and y, $dy/ds = y'$, and s is used as the independent variable rather than t, which is allowed since velocity is constant in the absence of any electric fields. For now, all particles that make up the beam are assumed to have the same momentum. That topic will be taken up in a later section of the text, "Dispersion". Therefore, both field-free regions, called "drifts" and dipoles of length L, have a "transfer matrix" which propagates a particle along the reference orbit, $ds \sim dz \sim L$, and which describes the particle as moving in a straight line. That motion has a simple matrix representation. It is assumed that the motions in x and y are independent. Complications alter these assumptions, and will be mentioned later since the text goes from the simplest case and adds complications only later.

$$
\begin{gathered}
x = x_o + x_o' L, \quad y = y_o + y_o' L \\
x' = x_o', \; y' = y_o' \\
\begin{pmatrix} x \\ x' \end{pmatrix} = \begin{pmatrix} 1 & L \\ 0 & 1 \end{pmatrix} \begin{pmatrix} x_o \\ x_o' \end{pmatrix}.
\end{gathered} \tag{4.2}
$$

The approximate action of a dipole on the initial beam was shown previously. More details will be explored later in the discussion on beam dispersion due to off-momentum particles having differing bend angles than the bend angle of particles with the nominal momentum.

In the case of the quadrupole, the equations of motion are found by deriving the magnetic fields from the scalar magnetic potential, using the Lorentz force equation, and using s as the independent variable rather than time. The resulting equations of motion are

$$
\begin{aligned}
B_x &= -(dB/dr)y \\
B_y &= -(dB/dr)x \\
d^2x/d^2s &= k_Q x, \quad d^2y/d^2s = -k_Q y \\
k_Q &= qe(dB/dr)/p, \quad \phi_Q = \sqrt{k_Q} L \\
1/f &\sim k_Q L = \sqrt{k_Q}\phi_Q.
\end{aligned} \tag{4.3}
$$

The equations of motion are those for simple harmonic motion with sinusoidal or hyperbolic solutions depending on the sign of k_Q. If the motion in the x plane is sinusoidal, that in the y plane it is hyperbolic, and vice versa. The parameter k_Q has the dimensions of inverse length squared. A useful way to express k_Q is $k_Q = qe(dB/dr)/p = (dB/dr)/B\rho$. A focal length, f, is defined as it is in classical optics. It is appropriate when the focusing effect is small.

There is no explicit off-momentum dependence of the focal properties of the quadrupole in the first order analysis. The simple approximate analysis, similar to what was done for the JPARC beam, shows that the focal length is as quoted in Eq. (4.3). The Lorentz force along x is proportional to x. There is a momentum impulse that results in a "kick" or change in angle.

$$\begin{aligned} F_x &= -qe\beta B_y = -qe\beta(dB/dr)x = k_Q\beta px \\ dt &\sim L/\beta, \quad dp_x/p = dx' = -k_Q Lx = -x/f. \end{aligned} \tag{4.4}$$

The force in the x direction may be focusing (Q_F). If so, the force in the y direction is defocusing (Q_D). Therefore, in contradistinction to classical optics, the minimum number of elements to provide overall focusing in both x and y is two or a doublet structure of both types of quadrupole elements.

These equations are linearized in terms of x, y, and s because the elements are idealized as a pure multipole. Only these "first order" effects are treated in this text. Higher order additions are treated in many advanced textbooks. That linearizing approximation allows the use of matrices. If the motion is focusing in the x direction, it must be defocusing in the y direction, and vice versa.

The harmonic solutions are sin and cos, while the diverging solutions are sinh and cosh. The matrix representation for passage through a quadrupole is

$$\begin{aligned} M_{Q_F} &= \begin{pmatrix} \cos\phi_Q & 1/\sqrt{k}\,\sin\phi_Q \\ -\sqrt{k}\,\sin\phi_Q & \cos\phi_Q \end{pmatrix} \\ M_{Q_D} &= \begin{pmatrix} \cosh\phi_Q & 1/\sqrt{k}\,\sinh\phi_Q \\ \sqrt{k}\,\sinh\phi_Q & \cosh\phi_Q \end{pmatrix}. \end{aligned} \tag{4.5}$$

The thin lens result is recovered in the case where ϕ_Q is very small: $1/f \sim k_Q L$. In the case of both dipoles and quadrupoles, the treatment here assumes no edge effects due to fringe fields. These effects clearly exist, but are ignored here in the interest of simplicity.

$$M_{Q_F} \to \begin{pmatrix} 1 & 0 \\ -1/f & 1 \end{pmatrix}, \quad M_{Q_D} \to \begin{pmatrix} 1 & 0 \\ 1/f & 1 \end{pmatrix}. \tag{4.6}$$

$$\text{Bx} = -\text{dBdr}\,y$$

$$\text{By} = -\text{dBdr}\,x$$

$$\text{MF} = \begin{pmatrix} \cos(\text{ph}) & \frac{\sin(\text{ph})}{\sqrt{k}} \\ -\sqrt{k}\,\sin(\text{ph}) & \cos(\text{ph}) \end{pmatrix}$$

$$\text{MD} = \begin{pmatrix} \cosh(\text{ph}) & \frac{\sinh(\text{ph})}{\sqrt{k}} \\ \sqrt{k}\,\sinh(\text{ph}) & \cosh(\text{ph}) \end{pmatrix}$$

Figure 4.3: Symbolic output from thc script "Quad_symbolic".

The script "Quad_symbolic" gives the fields in a quadrupole, the focusing matrix M_F and the defocusing matrix M_D in the other plane, and finally plots the quadrupole potential and fields. The fields are derived symbolically from the potential using the utility "diff" while the potential plot uses "contour", and the fields are derived numerically from the potential using "gradient" and displayed using "quiver". The symbolic output appears in Figure 4.3, while the potential and fields appear in Figure 4.4. The script is a "live" file in this case, so one can follow the calculations step by step as desired.

Numerically, for example, assuming a quadrupole of length $L = 3\,\text{m}$ with gradient $dB/dr = 10\,\text{T/m}$, the square root of k_Q is $0.17\,\text{m}^{-1}$, the quadrupole angle is $\phi_Q = 0.51$ rad, and the focal length, f, for a particle with $p = 0.1$ TeV is 11.2 m. In general for beams and accelerators the quadrupole focal length is larger than the physical quadrupole magnet length. Stronger focusing should be avoided if it is not necessary for financial reasons.

The script "Quadrupole_Thick_Thin" explores the differences between the full quadrupole orbits and the thin lens approximation. Comparisons are made for both focusing and defocusing quadrupoles

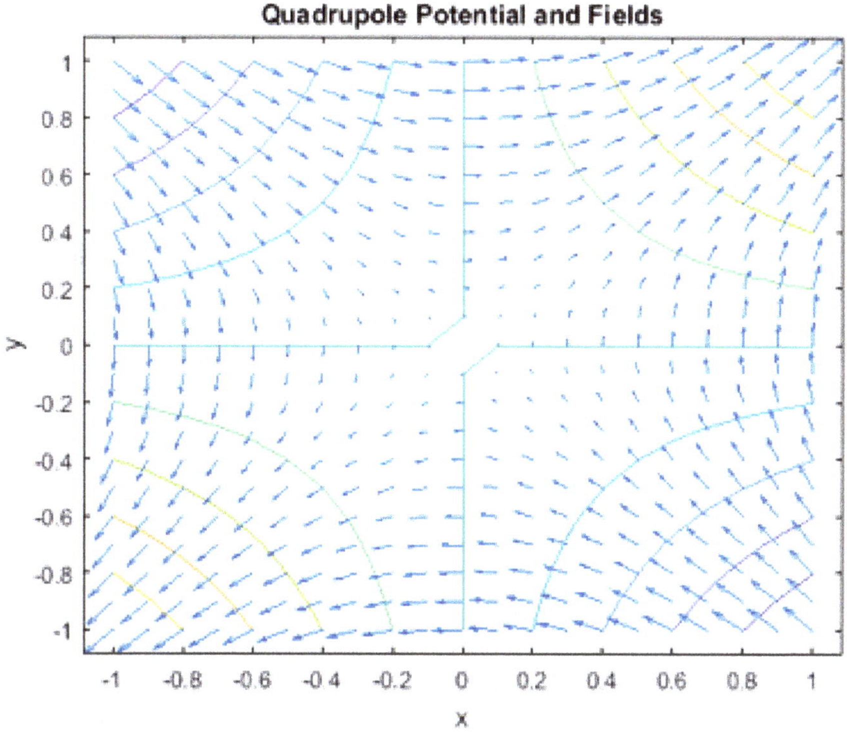

Figure 4.4: Equipotential lines for the quadrupole potential. The magnetic fields are derived from the potential and are displayed using "quiver" with arrows of length proportional to the field magnitude and of direction that of the field. The origin is field-free by construction.

and for both position and angle. The printout of the script appears in Figure 4.5, while the plots for a focusing quadrupole in position and angle appear in Figure 4.6.

The thick quadrupole is 6 m long and has been tracked in 1 m increments in length. The thin lens' position only changes at the quadrupole centerline. The thick lens' position changes continuously. Nevertheless, even with a quite thick lens, having a quadrupole angle of 58.4°, the exit positions agree quite well. A similar situation obtains for the angles. The thin lens shows an impulse in angle at the centerline, while the thick lens displays continuous evolution of the angle. Again, at the quadrupole exit the agreement of thick and thin is not very poor.

```
>> Quadrupole_Thick_Thin
  compare thick and thin quadrupole orbits F and D


 Quadrupole, L = 6 m, dB/dr = 10 T/m, p = 100 GeV
 phiQ = 1.02 rad, f = 5.75 m
```

Figure 4.5: The script printout which defines the quadrupole parameters in the example.

In the rest of the text, extensive use will be made of the thin lens approximation because of these agreements and the simplicity afforded by the approximation.

It is still worth keeping in mind that in most applications the quadrupoles are quite weak. However, they need not be. When the value of ϕ_Q is increased tenfold in the "Quadrupole_Thin_Thick" script the angular comparison of a thin and a thick orbit becomes very different, as seen in Figure 4.7. In this script the quadrupole orbit is sampled at only six locations in the magnet, so the smooth orbit looks somewhat jagged.

An "app" is available for exploring the full orbit through a quadrupole. The text box explains the fixed parameters. A switch can be used by the user to select a focusing or defocusing quadrupole. The incident values of the position and angle are chosen and the start button displays the orbit in x and x' through 20 steps of the track through the quadrupoles. Starting values are shown in Figure 4.8. The user has full freedom to set the initial incoming ray, and changing the "slider" values is a useful way to acquire an intuition about the resulting orbit.

The simplest system with focal properties in one transverse dimension is a quadrupole [Eq. (4.3)] with drift spaces [Eq. (4.2)] of length d_1 before the quadrupole and d_2 afterward. The successive passage of a particle through the three elements is accomplished by matrix multiplication. In MATLAB this can be done rapidly using command line inputs, as shown in Figure 4.9.

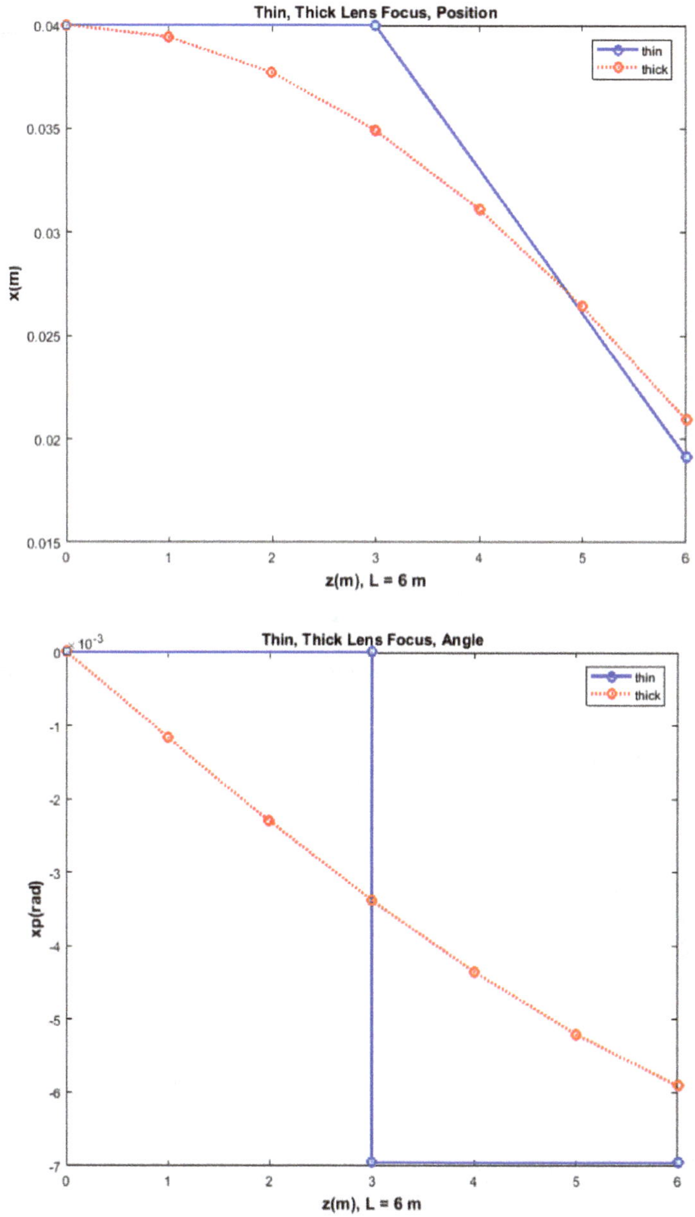

Figure 4.6: Trajectories of an initially parallel ray incident on a thin and a thick quadrupole.

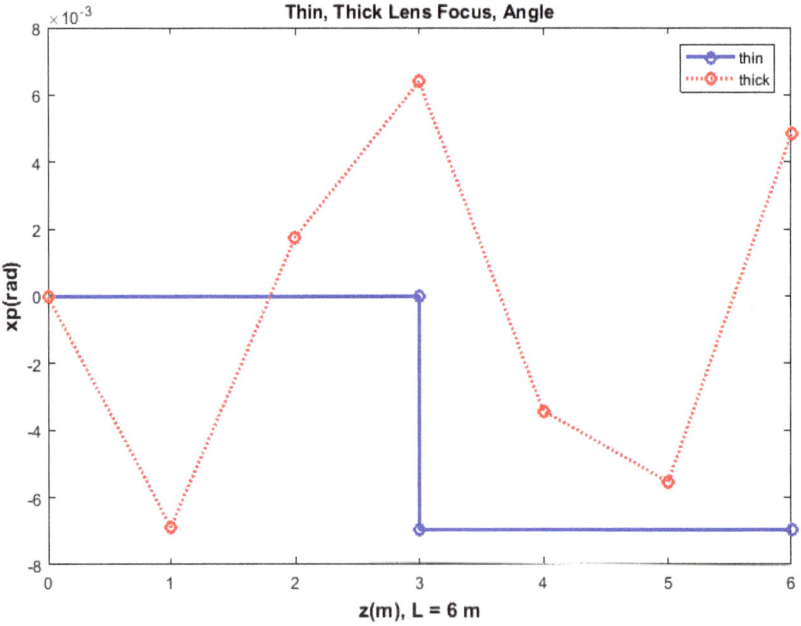

Figure 4.7: Angular orbit through a thick lens with a ten fold increase in the quadrupole angle with respect to the results shown in Figure 4.6.

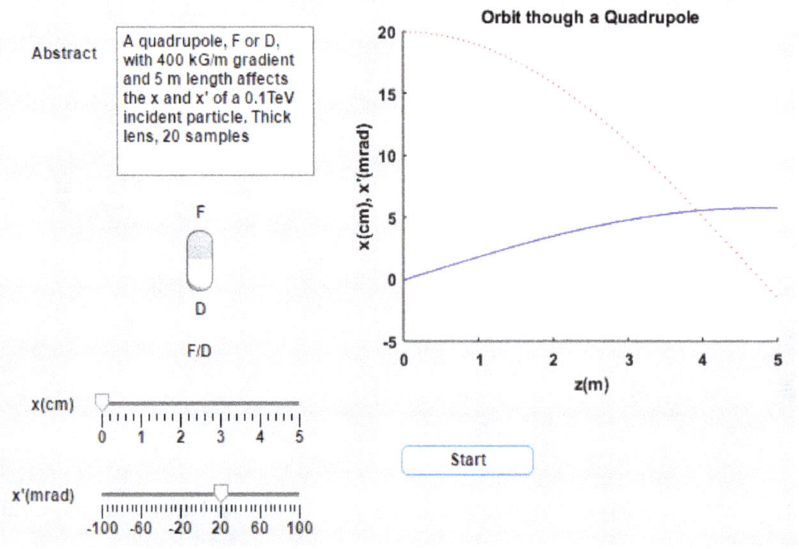

Figure 4.8: Output from the app "Quad_Orbits". The solid line is the position, while the dotted line is the angle.

```
>> syms d1 d2 f
>> M1=[1 d1 ; 0 1]; Mq=[1 0 ; -1/f 1]; M2=[1 d2 ; 0 1]; MT=M2*Mq*M1;
>> MT

MT =

[ 1 - d2/f, d2 - d1*(d2/f - 1)]
[     -1/f,           1 - d1/f]
```

Figure 4.9: Symbolic matrices for the single quadrupole plus two-drift-space system.

If the initial x is x_o and divergence is x'_o, then an initial pointlike beam has $x_o = 0$. A final pointlike beam is obtained if MT(1, 2) = 0, while a parallel beam occurs if MT(2, 2) = 0. An initially parallel beam, $x'_o = 0$, is focused to a point if M(1, 1) = 0. The last two conditions are met when $d_1 = f$ and $d_2 = f$ respectively, as seen by looking at Figure 4.9. The point-to-point focus occurs when $1/d_1 + 1/d_2 = 1/f$, an equation well known in classical optics. It is now clearer, in analogy with classical optics, why f is called the focal length.

Focus in one plane, say $(x,\ z)$, means defocusing in the other, $(y,\ z)$. Therefore, a minimum of two quadrupoles is needed to make the beam focusing in both transverse dimensions. That consideration brings us to the doublet of quadrupoles and an examination of the conditions for overall focusing in both transverse dimensions. This configuration is the next-simplest one following a single quadrupole and the simplest that admits of an overall focusing solution. The doublet is a very common configuration in both beams and accelerators.

Chapter 5

Quadrupole Doublet, Triplet

> My interest in matters more directly concerned with the handling of particles was growing, in the meantime, stimulated by many contacts with people understanding accelerators.
>
> — **Simon van der Meer**

The individual building blocks of beam and accelerator systems have now been defined at least in the lowest order of approximation. The idea is to start as simply as possible, build up an intuition by exploring simple systems, and later add complexity. The vacuum in which the particles move is assumed to be perfect. The beam particles all have the same momentum and do not interact except with the externally generated fields. The drift spaces are simple. The dipoles are "sector" dipoles (defined later) with no fringe fields, so they act like drift spaces. The quadrupoles are thin lenses and have no chromatic effects. The elements are all perfectly manufactured and perfectly aligned to the reference trajectory, which is in the $(x,\, z)$ plane. Since quadrupoles can be focusing or defocusing, a convention is adopted where F refers to x focusing (y defocusing) and D refers to x defocusing.

The script "symbolic_doublet" calculates the symbolic transport matrices through a system consisting of a drift space, a focusing quadrupole Q_F, a second drift space, followed by a defocusing quadrupole Q_D, and finally a drift space to a location where a constraint is placed on the overall transport matrix.

Numerically the distances of the system are defined by the user input in the command window, and then the focal length, thin

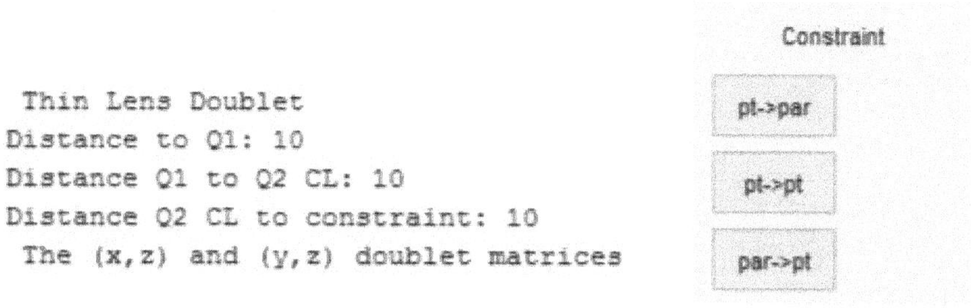

Figure 5.1: Dialogue for the script "symbolic_doublet". The geometry is user defined and the focal condition is chosen using the "menu" utility.

lens, of the two quadrupoles are solved for symbolically using the MATLAB utility "solve". Having defined the numerical values of d_o, the first drift space, d, the distance between the centers of the two quadrupoles and d_c, the distance from Q_D to the constraint location, the numerical solution is computed using the utility "eval". The focal condition is user-chosen from a popup menu. A sample dialogue printout appears in Figure 5.1.

In this example $d_o = d = d_c = 10\,\mathrm{m}$. The choice of focal condition is point -> parallel [$\mathrm{M}(2,2) = 0$], point -> point [$\mathrm{M}(1,2) = 0$], and parallel -> point [$\mathrm{M}(1,1) = 0$]. The condition for parallel -> parallel is not possible for a doublet for the two transverse dimensions simultaneously. A plot of the symbolic results for the full x and y transport matrices is given in Figure 5.2. The solution for the two focal lengths is shown in Figure 5.3.

In the special case where Q_F and Q_D have the same focal lengths, there is a net focusing effect, and $\mathrm{M}(2,1) < 0$, for both x and y motion with $\mathrm{M}(2,1) = -d/f^2$ giving an effective doublet focal length of f^2/d. A ray trace for point-to-point focusing appears in Figure 5.4.

The user is encouraged to "play" with the focal conditions and with the input geometry. This is the simplest focal system for the two transverse dimensions and an intuition can be built up by trying a variety of focus modes and doublet geometries.

In a more realistic situation the script "Doublet_Driver" treats the case of a thick lens doublet. The geometry is again a user-supplied input, as is the choice of focal condition. The starting values for a numerical fit are obtained from the thin lens initial

```
 Thin Lens Doublet
Distance to Q1: 10
Distance Q1 to Q2 CL: 10
Distance Q2 CL to constraint: 10
 The (x,z) and (y,z) doublet matrices
/             / dc     \               /             / dc     \     \                 \
|      dc + d | -- + 1 |               |      dc + d | -- + 1 |     |                 |
| dc          \ f2     /               | dc          \ f2     /     |     / dc     \  |
| -- - ----------------- + 1, dc + do  | -- - ----------------- + 1 | + d | -- + 1 |  |
| f2           f1                      \ f2           f1            /     \ f2     /  |
|                                                                                     |
|               d                                   / d           \                   |
|              -- + 1                               | -- + 1      |                   |
|           1  f2                            d      | f2        1 |                   |
|         -- - ------,                      -- - do | ------ - -- | + 1               |
\         f2     f1                         f2      \   f1     f2 /                   /

/        / dc     \                    /        / dc     \          \                 \
| dc - d | -- - 1 |                    | dc - d | -- - 1 |          |                 |
|        \ f2     /   dc               |        \ f2     /   dc     |     / dc     \  |
| ----------------- - -- + 1, dc + do  | ----------------- - -- + 1 | - d | -- - 1 |  |
|         f1          f2               \         f1          f2     /     \ f2     /  |
|                                                                                     |
|           d                                      / d           \                    |
|          -- - 1                                  | -- - 1      |                    |
|          f2        1                             | f2        1 |    d               |
|        - ------ - --,                     1 - do | ------ + -- | - --               |
\            f1     f2                             \   f1     f2 /   f2               /
```

Figure 5.2: Sample output from the script "symbolic doublet", showing the symbolic transfer matrices for both x and y.

approximations which were already imposed. The dialogue appears in Figure 5.5.

The MATLAB script uses the numerical utility "fminsearch" to simultaneously impose the nonlinear constraints on the x and y matrix elements. The driver script calls "Doublet_Thick_Thin", which first calls "Doublet_Thin" to obtain initial approximate thin lens solutions. Then with a choice of focal conditions made by the use. The utility "fminsearch" is used by "Doublet_Fit" to minimize elements of the transfer matrix in both x and y. The full quadrupole matrix elements are found using "Quad_Matrix". Finally, "Doublet_Plot" follows the beam at 100 points along z to the focal location, so that rays in the thick quadrupoles are accurately represented, in contrast to the case with the thin lens approximation. Rays for the thin lens are provided for comparison purposes. The rays for both the thin and the thick lens appear in Figure 5.6.

```
Symbolic Solution for f1 , f2
                         /         d (d + do)          \
do (d + dc) sqrt| ------------------------ |
                         \ (d + dc) (d + dc + do) /
------------------------------------------
                              d + do

                /         d (d + do)          \
dc sqrt| ------------------------ |
                \ (d + dc) (d + dc + do) /
```

Figure 5.3: Symbolic solution for the Q_F and Q_D focal lengths.

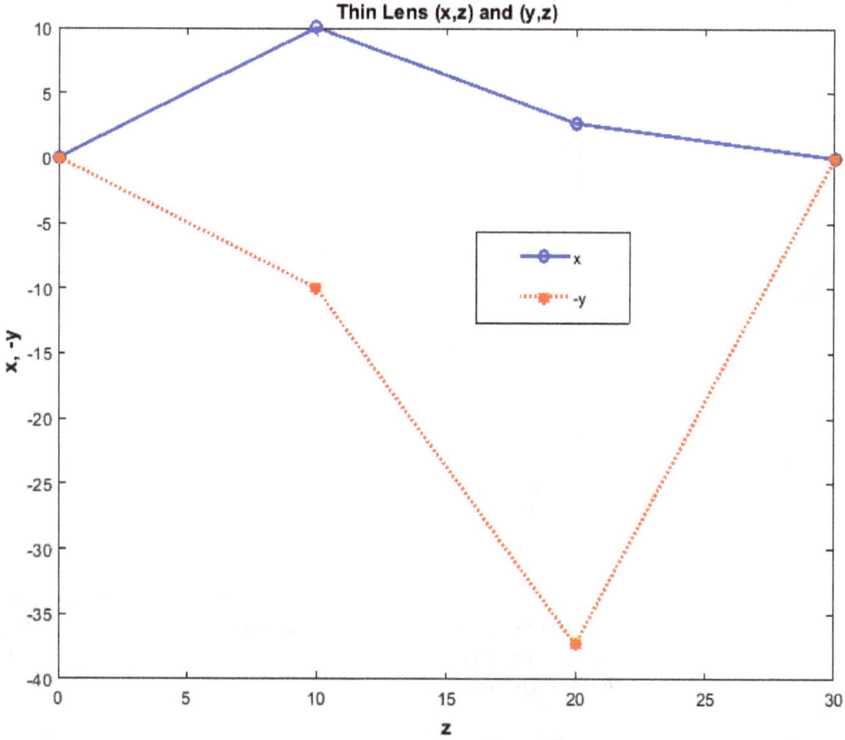

Figure 5.4: Rays for a thin lens doublet in the point-to-point condition for motion in both x(o, solid blue) and y(*, dotted red).

```
>> Doublet_Driver
 Quadrupole Doublet - Thick and Thin Lens
Type,1 pt -> par,2 pt -> pt,3 par -> pt,4 par -> par
1
 Example Z = [7.5, 12.5, 17.5, 22.5, 30.]
z Locations of Q1in, Q1out, Q2in, Q2out, Constraint[ ]
[8 13 18 23 30]
 Focal Lengths-Thin Lens, fx 7.33352 , fy 14.3178 Focal Lengths, fx 6.2674 , fy 12.6518,
```

Figure 5.5: Output of the dialogue with the script "Doublet_Driver".

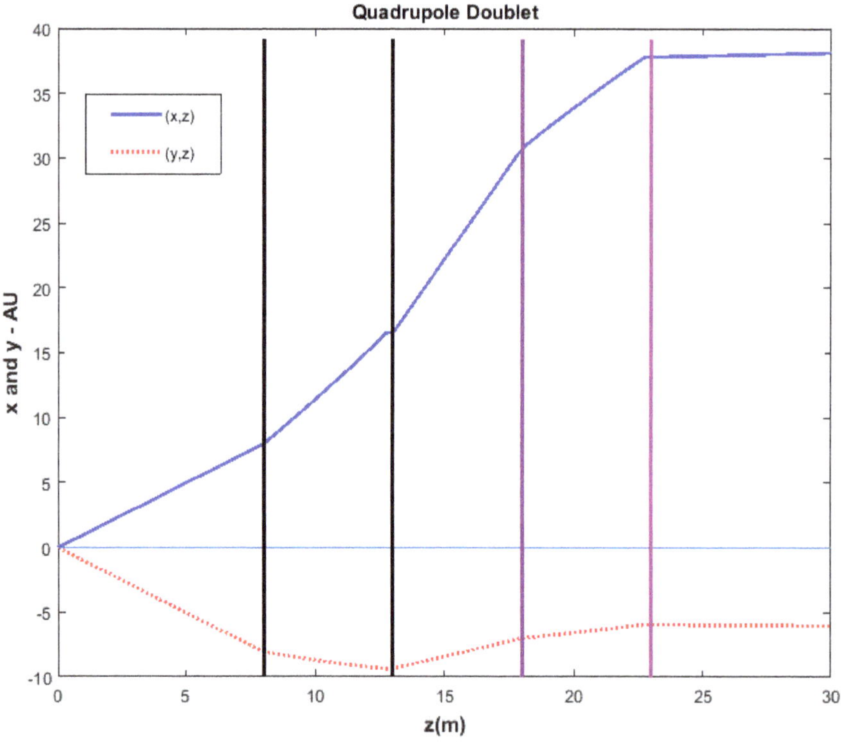

Figure 5.6: Rays for the point-to-parallel condition for a thick lens quadrupole doublet. Note that in x the first quadrupole is now x-defocusing. A doublet has a rather asymmetric final beam size. The boundaries of the two quadrupoles are indicated by the locations of the vertical lines.

It is of interest to note that the focal lengths in the thin lens approximation are larger than those of the thick lens calculation. That is because the change in x' occurs at a point in z in the thin lens approximation but is spread in z over the thick lens, thus

```
thin lense symmetric - symbolic

 Thin Lens Triplet
Distance to Q1: 10
Distance Q1 to Q2 CL = Distance Q2 to Q3 CL: 10
Distance Q3 CL to constraint: 20
```

Figure 5.7: A specific user dialogue for the symbolic solution of the symmetric thin lens triplet. The "menu" choice now has parallel-to-parallel added since this can be accomplished in a triplet as opposed to the doublet.

softening the maximum excursion to a less "hard" focus. However, the differences are typically not great and the thin lens solutions provide a good starting point for most solutions. The scripts that follow will largely use only thin lens estimates. The differences from a full thick lens evaluation have been shown now for both a single quadrupole and a quadrupole doublet. These examples supply the necessary input for deriving an intuition about the validity of the thin lens approximation.

The next-most-complex set of quadrupoles is the triplet. The symmetric triplet, where the two focusing quadrupoles have the same f value, f_1, while the defocusing one has a second value, f_2, is solved for using the MATLAB symbolic tool "solve" in thin lens approximation. The quadrupoles are assumed to be equally spaced. The user dialogue appears in Figure 5.7. In this case CL means the centerline of the quadrupole where the angles are changed. Often the triplet has the center quadrupole at double strength or double length. This latter configuration tends to yield beams with roughly equal excursions in the x and y planes. Such "round" beams are often desirable in experimental conditions for targeting external beams or in colliding beam applications.

A specific output for the point-to-point constraint for $d_o = 10\,\mathrm{m}$, $d = 10\,\mathrm{m}$, and $d_c = 20\,\mathrm{m}$ appears in Figure 5.8. Note that the x and y rays are more symmetric than in the case of the doublet configuration. The focal lengths are 9.5 m and 7.8 m for the F and D quadrupoles respectively, also rather more symmetric than in the doublet case. The symbolic expressions for the transfer matrices and even for the focal length solutions are not particularly illuminating

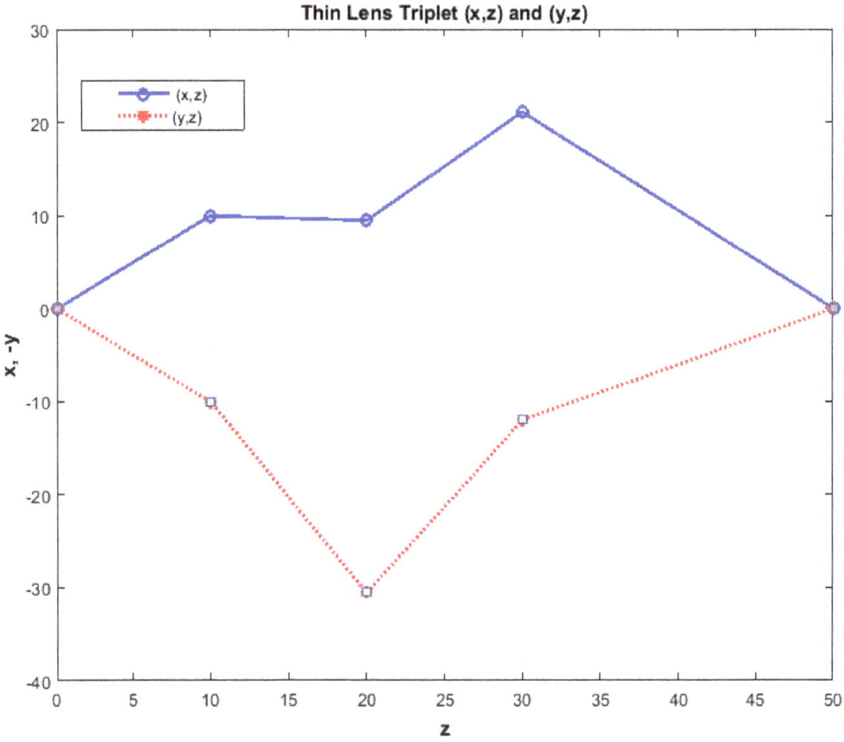

Figure 5.8: Rays in x(o, solid) and y(*, dotted) for the symmetric doublet in the point-to-point configuration. The maximum x and y excursions are now approximately equal.

and, therefore, are not shown here. The user can examine them if this is desired using the command window without the ; symbol, which suppresses printout.

One major application of triplets is in colliding beams. In that case, the reaction rate depends on minimizing the beam size simultaneously in the two transverse dimensions. The triplet accomplishes this and often appears in "low beta" designs at colliding beam facilities. This use of triplets will be discussed later.

One interesting fact arises sometimes in the triplet solutions. Occasionally, a symbolic solution is not possible for MATLAB. The utility "solve" which is used notes the inability to solve the problem symbolically and attempts a numerical solution instead. This aspect

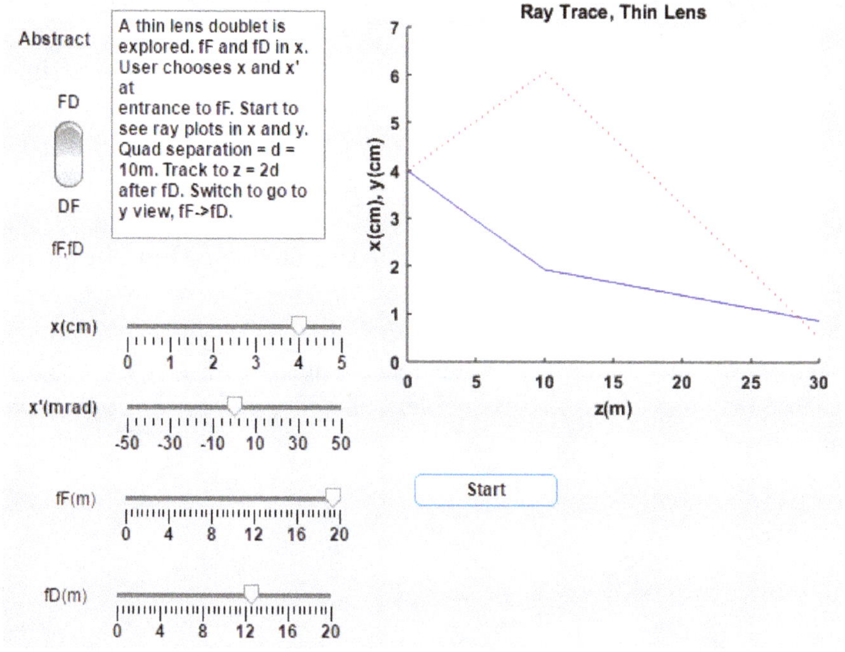

Figure 5.9: Output of the app "Doublet_General" for a particular choice of x, x', f_F, and f_D.

of "solve" can be very useful when the magnet configurations become more complex.

The doublet remains the fundamental building block because it is the simplest configuration that achieves a focus in both transverse dimensions. Because of that, an "app" was written which allows the user to vary many parameters of a doublet system in the thin lens approximation.

Ray traces in both x and y are displayed. Only two values are fixed: the distance between quadrupole centerlines, $d = 10\,\text{m}$, and the total range over which the rays are traced, $3d = 30\,\text{m}$. The user has a "slider" to choose the x and x' values at the entrance to the first quadrupole. These settings also force y and y' to be the same values. Two other sliders make the focal lengths of the first and second quadrupole user-defined values. Finally, a switch changes the polarity of the quadrupoles; F followed by D or D followed by F.

After the choice of the slider values, the rays are traced by pushing the "start" button.

Many configurations can be selected. A particular set of settings yields the rays shown in Figure 5.9. This is a simple parallel-to-point setting, one which was already shown. The user is encouraged to "play" and see what can be found out about the doublet configuration.

A lot of time has been spent on the doublet. One reason for this emphasis is its ubiquity in beam designs. Another is that the unit cell of many accelerators is the "FODO", which contains a doublet of quadrupoles of equal focal length. The accelerator can then be thought of as many doublets pasted together in a repetitive structure.

Chapter 6

Beams

When I was 16 years old, I assembled a 2.3 million electron volt beta particle accelerator. I went to Westinghouse, I got 400 pounds of translator steel, 22 miles of copper wire, and I assembled a 6-kilowatt, 2.3 million electron accelerator in the garage.

— **Michio Kaku**

The elements of dipole, or "bending", magnets and quadrupoles can be combined into a beamline array. Particles make a single pass through this array, which makes them simpler than accelerators or storage rings, which must be stable over many passes or "turns" through an array. Therefore, beams are considered first, followed by discussion of accelerators and storage rings.

The creation of an external beam begins with the extraction of a "primary proton" from an accelerator, for example. Primary electrons are also available from circular or linear electron accelerators. The details of how this is accomplished, slowly or rapidly, are not discussed here. To be concrete, imagine proton extraction from the Fermilab Main Injector (MI). This extracted beam may be transported directly to an experiment.

Another possibility is the transport of the extracted proton beam to a target where the protons interact and produce a variety of secondary particles. Examples of secondary particles are the strongly interacting pions or kaons, π, K, already mentioned in the JPARC example. Charged secondary particles can then be transported to an experiment by a magnetic array. The target can typically be considered a point source, so a doublet or triplet can capture the diverging beam and transport it to an experiment where it may again be focused with a triplet, for example, if a round beam is desired.

Tertiary beams are also possible. They can be formed using the decay products of π or K secondary beams. Both types of secondary particles decay into muons, electrons, and neutrinos. The muons can be enriched by utilizing the fact that they approximately only lose energy by ionization while the strongly interacting beam components can be removed by interposing large amounts of material. A very simplified example is given later in this chapter. At Fermilab, examples are the muon beams used in the g-2 or Mu2e experiments.

Finally, neutrinos can be selected by the fact that they have no charge and interact very weakly. By interposing enough material, the muons can be removed since they lose energy by ionizing the material traversed, leaving only neutrinos. At Fermilab, an example is the neutrino beam to be sent to the DUNE experiment in a direct line through the Earth.

An example of a secondary beam is the low energy electrostatically separated kaon beam at JPARC. A ray trace of the beamline is shown in Figure 6.1. The use of crossed electric and magnetic fields was discussed previously. The target is taken to be point-like and a quadrupole doublet captures the beam. Bending magnets, labeled B, steer the beam and provide momentum selection.

A point-like focus in y is used to clean up the beam. A second quadrupole doublet makes the beam parallel in y so that the small angular difference between pions and kaons induced by the separator is not obliterated by the angular divergence of the beam. After the separator a doublet focuses at the "mass slit", where pions are removed by collimation. Yet another doublet brings the beam to a final focus for experimentation. In the figure, dipoles are labelled B, quadrupoles Q, and sextupoles S.

A new beamline element appears in the JPARC beam: the sextupole. It is required because of the dispersion in the beam caused by the fact that all beam particles, although momentum-selected using dipole, or bending, magnets and collimation, do not have the same momentum. In turn, this means that the separation in velocity

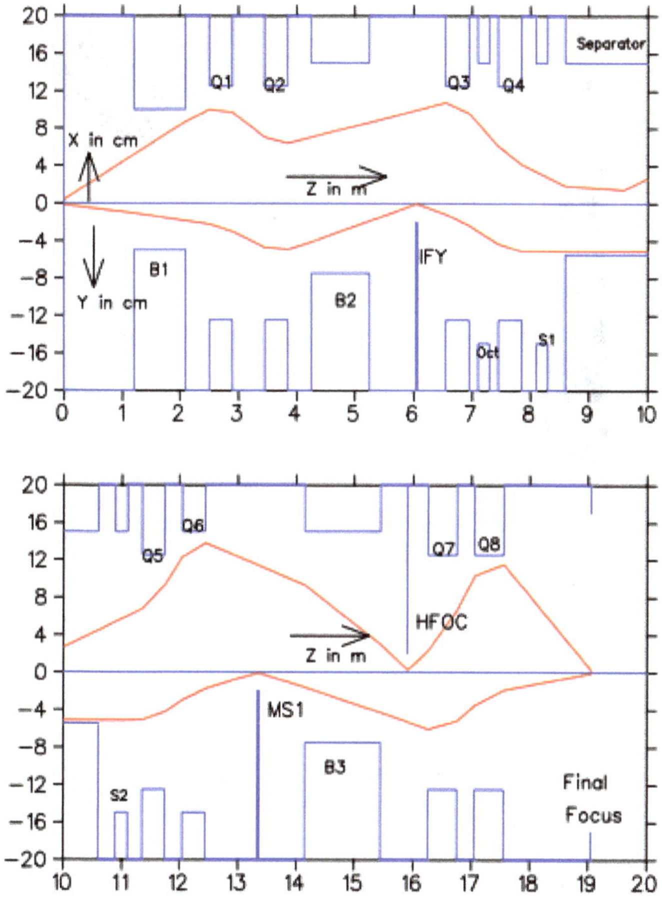

Figure 6.1: Rays in the JPARC separated beam. Note the presence of the sextupoles, S1 and S2. There is a vertical focus at 6 m, a parallel region at the separator, a vertical mass slit (MS) at 13.5 m, a horizontal focus at 16 m, and a final focus at 19 m.

of pions and kaons is smeared out by the momentum dispersion. The sextupoles are magnets provided to correct for this dispersion and thus sharpen the pion and kaon images at the mass slit.

The fields of dipole, quadrupole, and sextupole magnets are assumed to be

$$\begin{aligned} B_x &= (dB/dr)y + (d^2B/d^2r)xy \\ B_y &= B_o + (dB/dr)x + (d^2B/d^2r)(x^2+y^2)/2. \end{aligned} \tag{6.1}$$

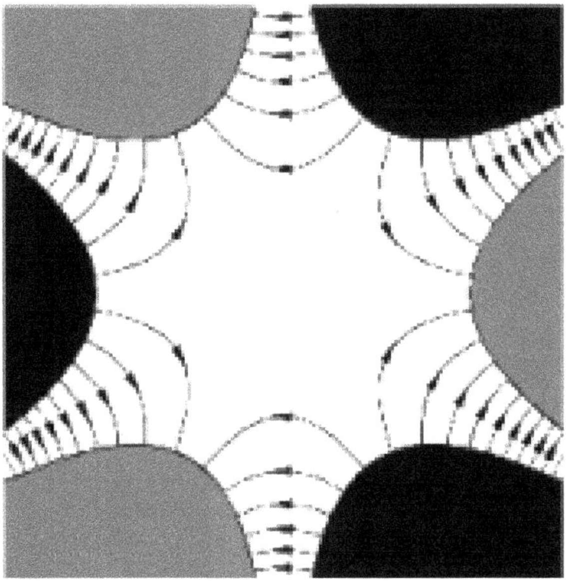

Figure 6.2: Field lines for a sextupole beam element. The field at the origin is zero, by construction. Off the axis, the field depends on the second derivative of the field with radius and is quadratic in the transverse coordinates.

That sextupole field is created by the arrangement of coils shown in Figure 6.2. They are derivable from a scalar potential, $\Phi_S \sim \cos(3\phi)r^3$.

The quadrupole provides a linear restoring force which leads to simple harmonic motion of the beam particles. The effect of a sextupole in the thin lens approximation is to impart an impulse which results in an effective focal length which is itself not a constant but is dependent on the x excursion off the beam centerline as shown, using Eq. (6.1) (x dependence only) in Eq. 6.2. Clearly, the sextupole is not a linear device. The force in the x direction is F_x, while the sextupole, of length L, "focal length", is f_S. Using the impulse approximation for the sextupole as was done for the quadrupole, a sextupole focal length can be defined.

$$\begin{aligned} &B_y \sim (d^2B/d^2r)x^2/2,\ F_x = -qe\beta B_y = dp_x/dt \\ &dp_x/p = dx',\ dx'/x \sim 1/f_S = (d^2B/d^2r)/\rho B(xL/2). \end{aligned} \tag{6.2}$$

An off-momentum but on-axis particle has a displacement, $x \sim D\delta$, where $\delta = dp/p$ and D is called the dispersion. The precise value of D depends on the specific layout of the dipole magnets in the specific case. The sextupole impulse, then, depends on $D\delta$, where D has the dimensions of length. The off-momentum particle experiences an effective focal length of the sextupole which depends in strength on dp/p, $1/f_S = (d^2B/d^2r)(LD\delta)/2B\rho$, which illustrates that the focusing power is proportional to the magnitude of the off-momentum value. The sextupole therefore directly corrects for the dispersion in the beam.

Since off-momentum particles have a centerline in a dipole which is displaced from the on-momentum centerline, assumed to be in the $(x,\ z)$ plane, the sextupole enables dispersion corrections. Aside from the use of sextupoles in dispersion correction in low energy beams, they are employed in corrections in most accelerators; for example, correcting for errors in the dipole or quadrupole components of actual magnets with their imperfections, such as off-axis nonuniformity. The forces used are quadratic in the transverse positions and therefore contribute nonlinear trajectories.

A sextupole beamline element appears in Figure 6.3, showing the conductor, the shaping steel, and the power leads.

Returning to a secondary beam as a mixture of pions and kaons, the two particle types may be identified in a high energy beam even if they are not physically selected as they are in the JPARC beam. One typical means of identification uses the Cerenkov effect, which is the emission of light when a particle traverses material in which it exceeds the velocity of light. This behavior is analogous to the existence of a bow wave when a boat exceeds the speed of wave propagation.

In the material, the velocity of light is c/n, where n is the index of refraction. The light is emitted at a specific angle, θ_c. A useful high energy approximation is

$$\begin{aligned} \cos(\theta_c) &= 1/n\beta \sim 1 - \theta_c^2/2 \sim 1/(1-\delta\beta)(1+\delta n) \\ \theta_c &\sim \sqrt{2(\delta n - \delta\beta)}. \end{aligned} \tag{6.3}$$

Figure 6.3: Photograph of a beamline sextupole magnet, showing coil, leads, and shaping iron.

The magnitude of the emission angle depends on the value by which δn exceeds $\delta\beta$. No light is emitted below the threshold for Cerenkov emission, $\beta = 1/n$. The emission angle is motivated using the script "Cerenkov_Threshold". Particles are propagated through a medium with n of 1.2 and emit outgoing circles of radiation, with user input shown in Figure 6.4. It is evident that constructive formation of a "Mach cone" occurs when the speed exceeds n. The user picks the velocity and watches a movie of the light emitted from different locations on the particle path. The light above the threshold appears in Figure 6.5.

Cerenkov counters are filled with transparent liquids or gas, and light is detected if a beam particle exceeds the threshold, $\beta = 1/n$. The counter can be used simply to decide if light was emitted, or the angle of the emitted light may be measured. The two types are called threshold and differential Cerenkov counters, respectively.

Some of the instrumentation set up in a beam created for detector testing at Fermilab is shown in Figure 6.6. The Cerenkov counters,

```
>> Cerenkov_Threshold
  Movie of the Cerenkov Mach cone


Input Velocity of Source w.r.t. Light, cos(thetac) = 1/beta*n: 0.7
 Below Cerenkov Threshold
Input Velocity of Source w.r.t. Light, cos(thetac) = 1/beta*n: 1.3
Cerenkov Angle = 50.1317 Degrees
```

Figure 6.4: User choices for the velocity of a particle in a medium with $n = 1.2$. Behavior both below and above the Cerenkov threshold can be explored.

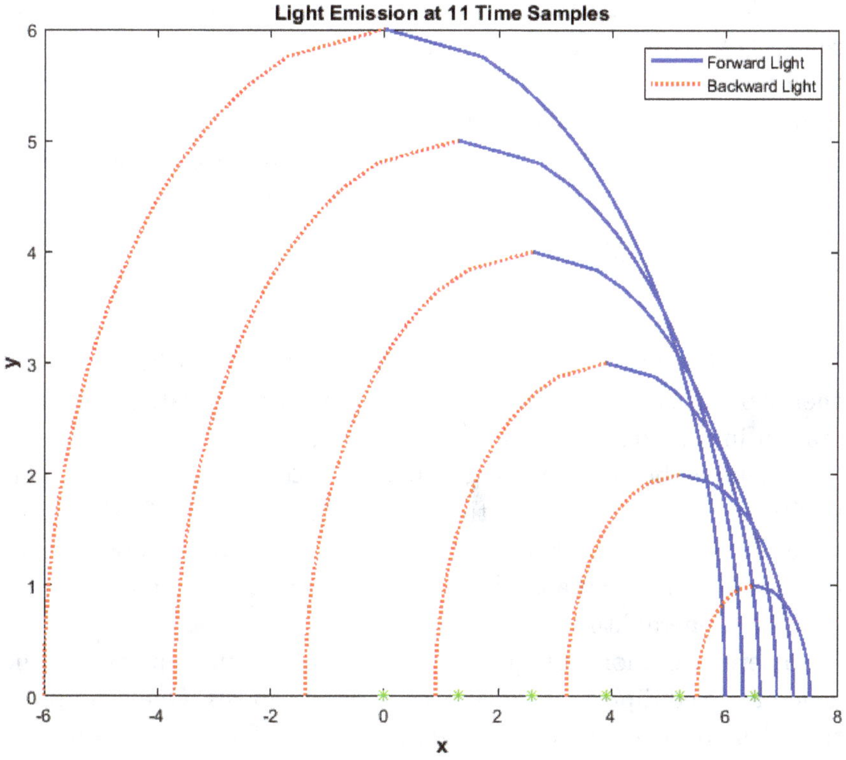

Figure 6.5: The last frame of a "movie" of light emitted by a particle moving with uniform velocity which is greater than the velocity of light in the medium. The outgoing waves reinforce in forming a "Mach" cone. The five discrete emission points appear as green *.

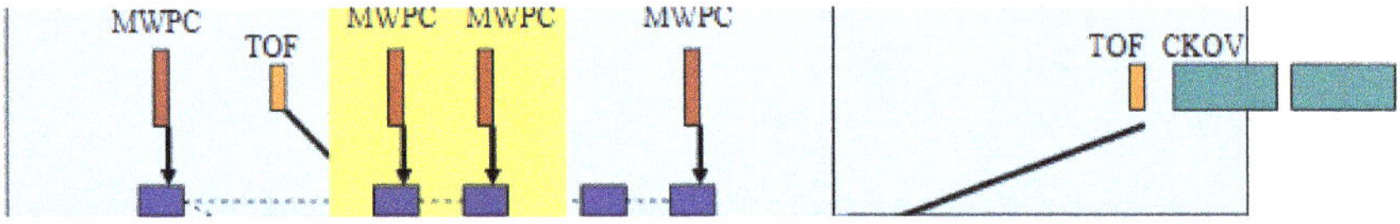

Figure 6.6: Instrumentation for the MT beamline at Fermilab. Incident particle examination employs MWPC for position measurements and TOF and Cerenkov counters for particle velocity selections at a fixed beam momentum.

CKOV, appear in green. There are also particle ionization detectors (labeled as MWPC) which allow the reconstruction of the trajectory of incident beam particles. Finally, there are scintillation counters called TOF (time of flight). Measuring the flight time between two fixed points separated by a distance L measures the velocity of the beam particle, which differs for, say, pions and kaons since they have the same beam momentum as defined by dipoles and collimators. The lighter pions go faster than the kaons.

$$\begin{aligned}\delta TOF &= (L/c)(1/\beta_\pi - 1/\beta_K) \\ &\sim (L/2cp^2)(m_\pi^2 - m_K^2).\end{aligned} \tag{6.4}$$

The TOF technique is one which is useful only with fairly low energy beams, as was the method of electrostatic separation. The separation in time deteriorates as the inverse square of the momentum.

A similar beam for testing apparatus exists at several locations at CERN, in particular at the H2 beamline of the CERN North Area. The details of CERN test beams can be found in documentation such as http://nahandbook.web.cern.ch/nahandbook/default/h2/1%20General.htm.

There are more unique elements that occur outside of the generally used dipoles, quadrupoles, and sextupoles. One example is the solenoid, which was mentioned previously. The transport matrix for a solenoid is shown in Figure 6.7. Because there is a uniform axial magnetic field, particles execute orbits which are circles in the (x, y) plane and straight lines in the (z, s) plane, as shown earlier (Figure 2.5). In Figure 6.7 the rotation of the transverse momenta by an angle θ is evident, as well as the effects of edge focusing at the entrance and exit of the solenoid.

Transport for (x, x', y, y')

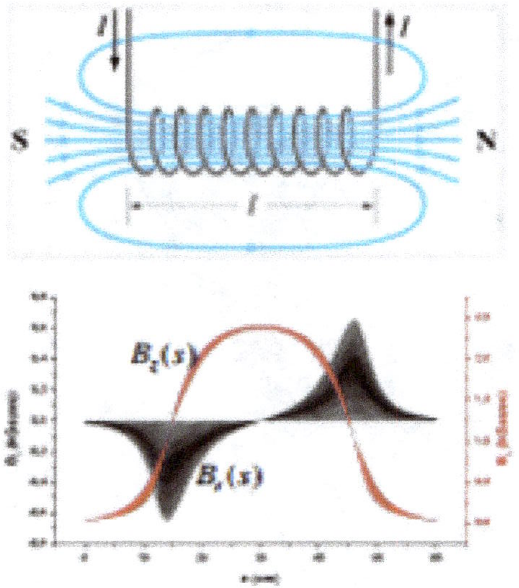

$$F_1 = \begin{pmatrix} 1 & 0 & 0 & 0 \\ 0 & 1 & K & 0 \\ 0 & 0 & 1 & 0 \\ -K & 0 & 0 & 1 \end{pmatrix} \qquad F_2 = \begin{pmatrix} 1 & 0 & 0 & 0 \\ 0 & 1 & -K & 0 \\ 0 & 0 & 1 & 0 \\ K & 0 & 0 & 1 \end{pmatrix}$$

$$M_B = \begin{pmatrix} 1 & \sin\theta/2K & 0 & (1-\cos\theta)/2K \\ 0 & \cos\theta & 0 & \sin\theta \\ 0 & -(1-\cos\theta)/2K & 1 & \sin\theta/2K \\ 0 & -\sin\theta & 0 & \cos\theta \end{pmatrix}$$

Figure 6.7: Fields in a solenoid and matrices for particle transport. The main bending occurs in M_B, while the fringe fields at the entrance and exit are represented by F_1 and F_2. The uniform field along the axis of the solenoid provides a capture trajectory for all particles moving transversely to that axis.

The solenoid provides a capture orbit for all particles moving transversely to the axis of the solenoid. Because of this solenoids are often employed to capture secondary beams. These solenoids are also very often the magnetic configuration of choice for detectors for colliding beams. Examples of experiments using solenoids are CDF and D0 at the Tevatron and ATLAS and CMS at the CERN LHC. Solenoids are also used in electron–positron detectors situated at SLAC and at the LEP facility at CERN.

The solenoid is also the magnetic element of preference in designing detectors for future facilities. A schematic of the solenoid employed in the CMS experiment at the LHC is shown in Figure 6.8. The detectors are placed inside the superconducting coil of the solenoid, except for the muon detectors. The coil is large enough to walk around inside. The solenoid provides a strong field of 4 T

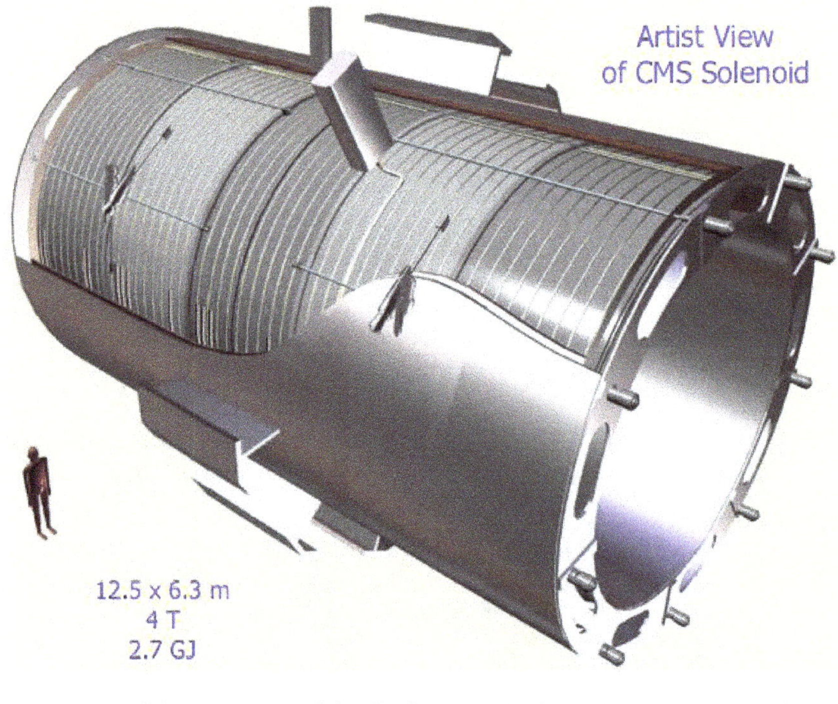

Figure 6.8: Schematic view of the CMS solenoid. The scale is set by the inclusion of a "standard"-sized person.

and has a stored energy of 2.7 GJ. It is presently the largest solenoid in the world by most measures.

Another specialized beam element has been used to provide efficient antiproton production for the Fermilab Tevatron Collider. A liquid metal is used to act as a lens caused by the current running transversely to the beamline and the resulting magnetic field, which focuses in the two transverse planes simultaneously. Extracted primary protons strike a nickel target and then are focused, point to parallel, for an 8 GeV antiproton, in a lithium lens (LIL), as displayed in Figure 6.9. The very large electric field in the liquid lithium is pulsed to provide maximum capture of secondary antiprotons. The current in the liquid lithium causes an azimuthal magnetic field which provides a focusing force for all transverse momentum orientations, $F_x \sim -Bp_z$, which supplies axially symmetric focusing.

More complex beams are also designed and modeled. For example, a simplified muon beam is explored in the script "Muon_Material_MC". The main script uses the MATLAB utility "rand" to generate random numbers for the simulation of a muon beam traversing a block of material. The material needs to be thick enough to remove all the strongly interacting particles from the beam, leaving only muons and neutrinos. Other subsidiary scripts are written as MATLAB "functions" to populate Gaussian distributions, "Gauss", to generate multiple Coulomb scattering for the muons in the material, "Mult", and to transform from the muon direction to the axis of the beam, "Euler". Using "functions" and invoking them from an overall but distinct script is very useful in making models

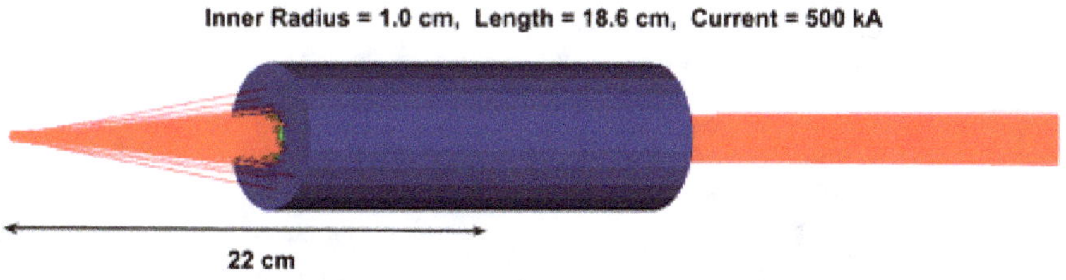

Figure 6.9: Schematic of the production and collection of antiprotons at Fermilab. The proton beam strikes a nickel target and the produced particles are collected by the field of the liquid metal lithium lens.

for more complex systems than are easily defined by a single, short "m" script.

The user chooses the incident muon momentum, the size of the incident parallel beam, and the length of the simulated block of user-defined material. The block is divided into many strips along the beam axis, so that the energy loss due to ionization and the multiple scattering angles can both be kept small and perturbative. Energy loss will be discussed in more detail in the section on medical applications of accelerators.

Muons are tracked through these strips and can lose all their energy, since their trajectory is not a straight line owing to the scattering. This simple Monte Carlo model should give a flavor of the complex models needed to accurately simulate complex beam configurations. The printout of a particular set of user choices appears in Figure 6.10. The script uses the MATLAB utilities "mean" and "std" to calculate the mean and standard deviations of an array, and "hist" to make a histogram of the values of an array.

The menu of user choices for materials appears in Figure 6.11 along with initial and final beam characteristics.

A movie of several trajectories for muons is supplied by the script. An example is displayed in Figure 6.12. The particle path is tracked through the 20 strips of material. Scattering and energy loss are evident in each strip. In this particular case the energy loss is large

```
>> Muon_Material_MC
  Program to make a model of a "parallel" muon beam passing through a block of
  material


Radiation Length, min dE/dx, Density and Muon Mass
Enter Mean Initial Muon Momentum (MeV), dp/p = 3%: 200
Enter Transverse sigma (cm): 2
Enter z Length of Block (cm): 100
Enter Number of Numerical Strips for Block of Fe: 20
Muon at the End of range
Muon at the End of range
```

Figure 6.10: Dialogue for the muon beam simple model. The user chooses the initial beam momentum and beam size along with the length of the block of material and the type of material. If a muon loses all its energy an end-of-range notice is printed.

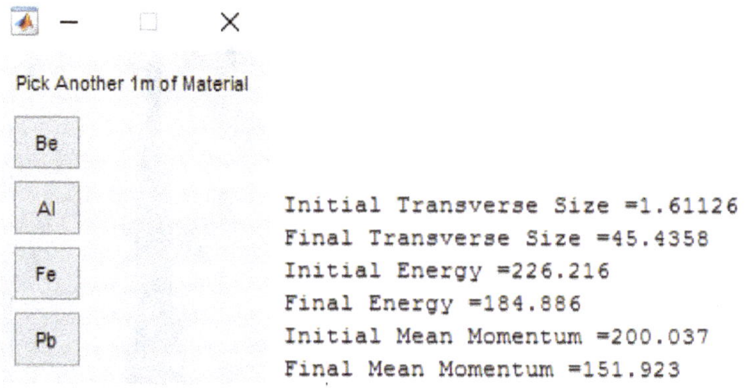

Figure 6.11: Menu for choice of the absorbing material and printout of the mean and standard deviations of the beam at the entrance and exit of the material.

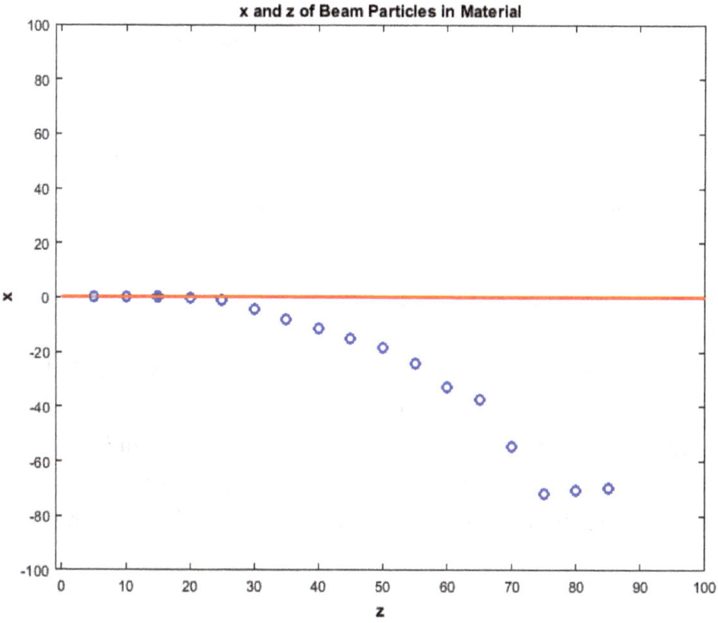

Figure 6.12: Movie frames of the path of a particular muon in material. The muon eventually loses all energy and stops in the iron.

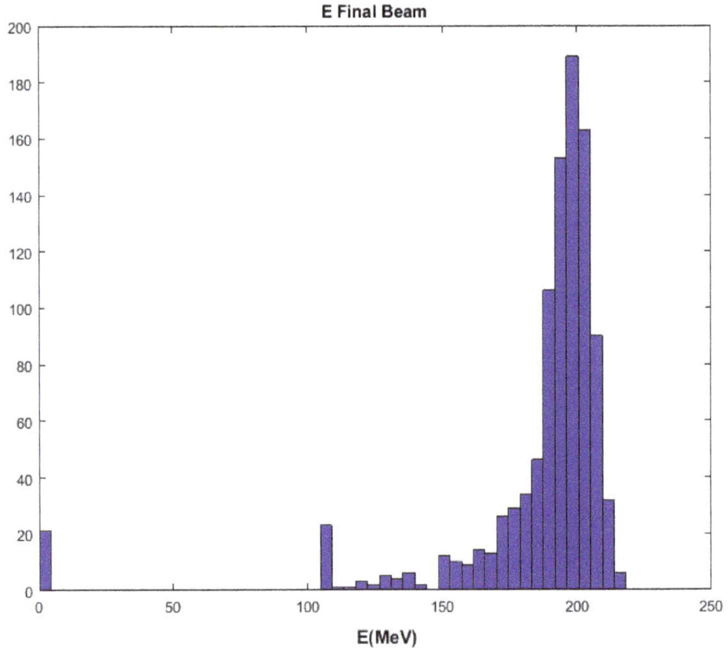

Figure 6.13: Energy of the muon beam particles at the exit of the block of material. The entries at energy of 0 indicate that the muon stopped, or came to the end of its range, before exiting the block having lost all its energy to ionization.

because the path length is large, and the muon actually stops in the material.

A histogram, using the utility "hist", of the energies of the muons exiting the 1 m block of iron appears in Figure 6.13. There are entries at zero energy, caused by muons losing all their energy. The distribution has a long "tail" at low energies, owing to the influence of scattering which increases the path length in the block in a stochastic fashion.

Chapter 7

FODO

Dr. Peter Venkman: "Why worry? Each of us is wearing an unlicensed nuclear accelerator on his back."
Dr. Ray Stantz: "Yeah. Well, let's get ready. Switch me on."

— **Ghostbusters**

For a beam to be viable there is only one pass that a particle needs to make through the elements of the beam. In the case of an accelerator, the particles captured must make many passes around a ring and have a stable orbit over many revolutions. This situation has necessitated a different formalism from the simple matrix theory of drifts, quadrupole, and dipole magnets that has been used so far although many of the concepts carry over. The treatment is not rigorous since there are many excellent textbooks on accelerators. Rather, the attempt made here is to give a sketch of the formalism and then illustrate some consequences in a variety of visual ways using the tools provided by MATLAB. This section concerns itself with the basic unit cell of an accelerator, called a "FODO".

A first look at the acceptance of a basic cell, a FODO, is explored using the script "FODO_Ellipse". The FODO is the basic building block of a repetitive structure which is essentially a series of basic doublets pasted together. For symmetry purposes, the FODO goes from the centerline of an F quadrupole through a drift distance d to a D quadrupole and then another drift of length d to an F quadrupole centerline. The FODO is reflection-symmetric. Such a structure might repeat as a basic element, or unit cell, of an accelerator ring. At this initial point the dipoles needed to steer the beam into a circular orbit are ignored. The focal lengths of the two quadrupoles are the same, making powering the magnets simple, in contrast to the beamline doublets, which normally had different focal lengths for the F and D quadrupoles. Output of the script appears in Figure 7.1. The user input defines the drift distance d and the focal length f.

```
>> FODO_Ellipse
  thin lens FODO doublet - symbolic


Distance from QF/2 to QD CL: 10
Common F and D focal length: 20
Transfer matrix for (x,dx/dz), (y,dy/dz)
/    2      2                  \
|   d   - 2 f   d (d + 2 f)  |
| - ---------,  -----------  |
|        2           f       |
|     2 f                    |
|                            |
|                   2     2  |
| d (d - 2 f)     d   - 2 f  |
| -----------,  - ---------  |
|       3                2   |
\    4 f              2 f    /

/     2      2                   \
|    d   - 2 f      d (d - 2 f)  |
|  - ---------,   - -----------  |
|         2              f       |
|      2 f                       |
|                                |
|                      2     2   |
|    d (d + 2 f)     d   - 2 f   |
| - -----------,   - ---------   |
|          3                2    |
\      4 f               2 f     /
```

Figure 7.1: Symbolic solution of the symmetric FODO unit cell transfer matrices in x and y.

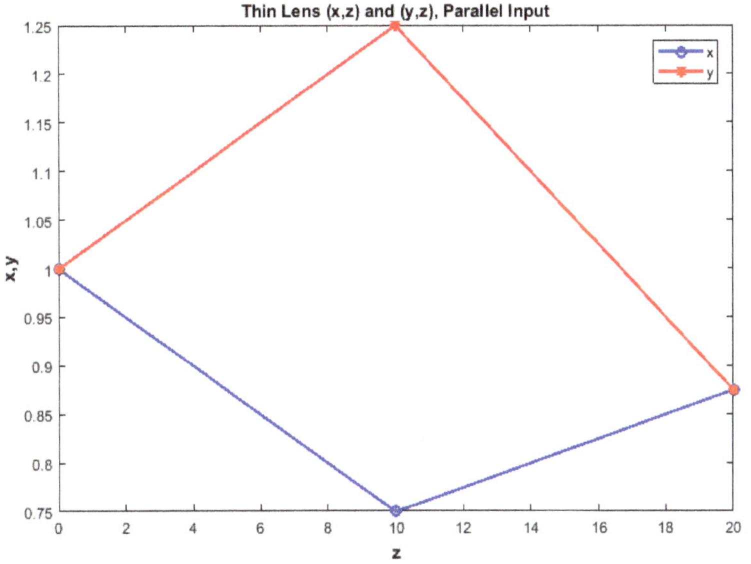

Figure 7.2: Thin lens ray trace for a FODO with $f/d = 2$. Both x and y have an overall focusing in traversing the FODO.

The symbolic transfer matrices for the FODO are shown in Figure 7.1. There is a net focusing of the FODO, M(2,1) < 0, if $f > d/2$, as seen in the figure, because M(2,1) is proportional to $d - 2f$. This is the basic principle of strong focusing accelerators. A plot of the thin lens result for the case where $f/d = 2$ is shown in Figure 7.2. The net focusing in both transverse planes is evident where the input parallel beam for both x and y of a size defined to be 1 is reduced to a size of 0.87 in both dimensions.

The acceptance of the FODO is examined by randomly generating rays in y and y', dy/ds, at the Q_F centerline and tracking them through the FODO. The aperture constraint is imposed by examining the ray location at the expected maximum location — the center of the D quadrupole [F in (y, y') by convention]. The MATLAB utility "rand" was used. It generates a random number when invoked, which is distributed uniformly between 0 and 1. Similarly, the x value of the limiting aperture is expected to occur at the center of the F quadrupole.

The resulting distribution of the rays which pass through the three quadrupole apertures appears in Figure 7.3. The distribution is shaped by the apertures. The apertures are of radius a in the quadrupoles and appear as straight lines at $y = -1$ and $+1$, taking $a = 1$, and at angles $+1/d$ and $-1/d$, since at the D center $|M(1,1)y_o + M(1,2)y_o'| < a$. The angled lines are the "shadow" of the apertures. The area enclosed by the lines is $4a^2/d$.

In the next chapter the generalization of Figure 7.3, with many apertures and many FODO traversals, will appear as the "beam ellipse". The net focusing condition can be further explored by supplying different f and d values to the script and examining the accepted phase space in $(y,\ y')$ and the ray traces.

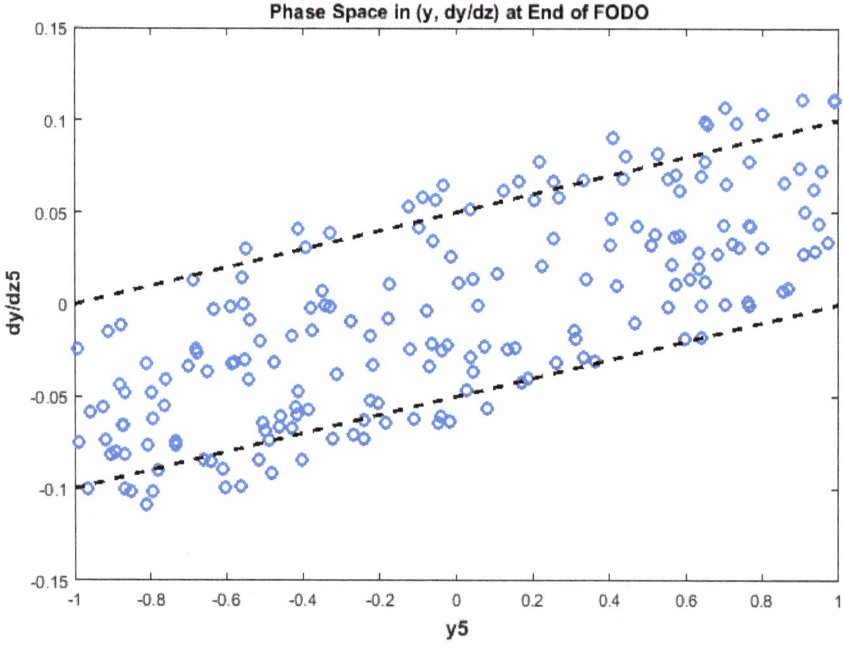

Figure 7.3: Rays in $(y,\ y')$ at the end of the FODO, the center of the F quadrupole, which pass through the other two quadrupole apertures at the centerline of the initial F quadrupole and the middle D quadrupole. The lines are the projection of the D aperture onto the final F quadrupole. The apertures are all taken to be of size 1 in a single dimension, or a slit.

In reality, the FODO consists largely of dipoles since the main purpose of the ring is to contain the largest energy beam in the defined circular vacuum pipe. The quadrupoles are there to capture and transport the beam. Fermilab was the first "separated function" machine, with distinct dipoles and quadrupoles. This led to the Fermilab logo, shown in Figure 7.4. The basic Main Ring (MR) FODO is shown in Figure 7.5.

Separate bending and focusing gives full flexibility to tune the gradients and the focal properties independently of the bending requirements. Previous accelerators, such as the Alternating

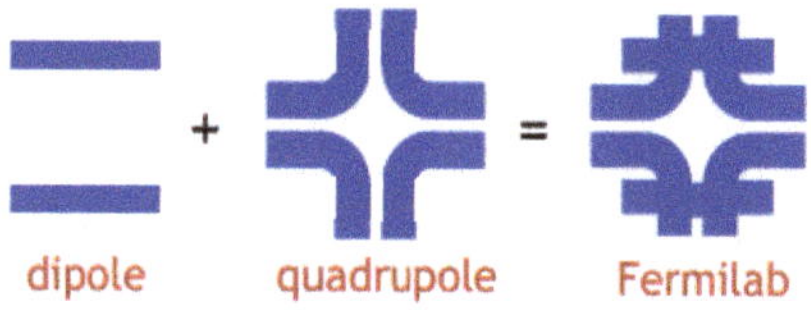

Figure 7.4: Fermilab logo showing the combination of separated dipoles and quadrupoles in the Main Ring.

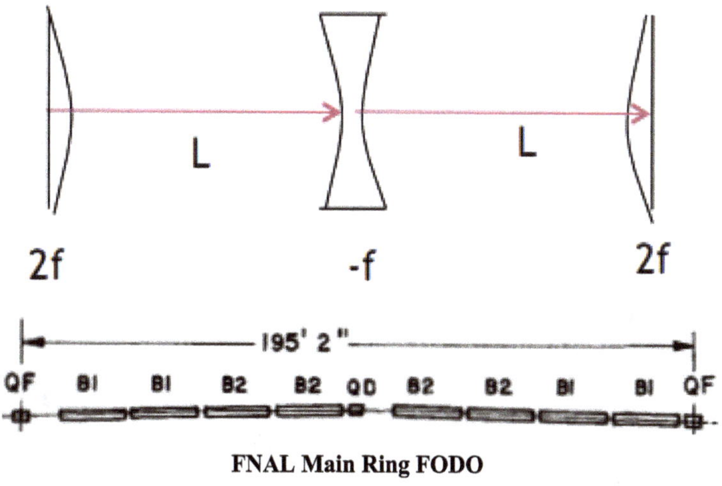

Figure 7.5: The Main Ring FODO. Note that the real estate is largely occupied by dipoles. The quadrupoles supply only the necessary focusing to keep the beam within the aperture of the vacuum pipe. Numerical values for the FODO elements are given in Appendix E and in Table 8.1.

Gradient Synchrotron (AGS) at Brookhaven National Laboratory (BNL), combined the functions in a fashion fixed by the shaping of the magnet pole tips. That combined functionality constrains the operation since focus and bend are coupled and depend on the pole tip hardware. Later machines have largely been separated function devices.

Chapter 8

Periodic Structures

Equipped with our five senses — along with telescopes and microscopes and mass spectrometers and seismographs and magnetometers and particle accelerators and detectors sensitive to the entire electromagnetic spectrum — we explore the universe around us and call the adventure science.

— **Edwin Powell Hubble**

An accelerator can be constructed by using "unit cells" such as the FODO, with dipoles to create the circular orbit. It is not the aim of this text to rigorously derive the equations followed in such a periodic structure but rather to explore their properties using MATLAB scripts that allow visualization and variation of the parameters of the particular problem at hand. Rigorous treatments are available in texts given in the references quoted at the end of this document.

MATLAB, being a matrix language, provides many tools for matrices; multiplication, *, inverse, inv(x), trace, trace(x), determinant, det(x), transpose ('), and several more, such as eigenvalues and eigenvector solution, eig. In addition, the utilities "solve" and "dsolve" allow for the symbolic solution of systems of linear and differential equations.

Since the magnetic forces considered in the FODO unit cell have no velocity-dependent dissipative terms to contend with, the matrix representing transport through the system has a unit determinant, $|M| = 1$, a condition familiar from classical mechanics. Under these conditions there is a conserved phase space, as stated in Liouville's theorem. This behavior is also familiar from classical-mechanical systems, such as planetary motion, where there is a conserved energy by which the orbits are defined. A closed orbit solution is assumed to exist which is periodic but is not "re-entrant" like the perihelion of Mercury since the forces are not inverse-square.

The eigenvalue equation for a system specified by a transfer matrix for a repetitive cell M is $M - \lambda I = 0$, where λ are the eigenvalues of the system and I is the unit matrix. This formulation is generally true in classical mechanics of conservative systems. As is the case in quantum mechanics, a state may be expanded in a complete set of eigenvectors. The number of eigenvalues equals the dimensionality of the system, which in this case of independent x and y motion is 2. For stable operation all eigenvalues must be complex and of unit magnitude.

The trace is the sum of the eigenvalues, and the determinant is the product, and both are invariant under a change of representation. In a diagonal basis, the eigenvector basis, for M where the eigenvalue equation is $\det(M) = [M(1,1) - \lambda][M(2,2) - \lambda] = 1$. This quadratic equation has a pair of complex roots, representing bounded periodic motion, if $M(1,1) + M(2,2)$ is less than 2. Therefore, the pair of complex eigenvalues, $\det(M) = 1 = \lambda_1\lambda_2$, of unit magnitude is $\lambda_{1,2} = e^{\pm i\mu}$, where μ is real and is called the tune or phase advance of the system. Unbounded motion is also possible, as it is in planetary motion for energies which are not sufficiently negative to bind the system. However, these solutions will not be explored.

For the eigenvalues to remain bounded over all time, or to be stable, the values of μ must be real, which requires

$$\mathrm{Tr}(M) = 2|\cos\mu| < 2. \tag{8.1}$$

The equations of motion admit periodic solutions with the periodicity of the unit cell and the circumference C. The equations for drifts, dipoles, and quadrupoles are all of the generic form:

$$\begin{aligned} &d^2x/d^2s + k(s)x = 0 \\ &k(s) = (1/B\rho)[\partial B(s)/\partial r]. \end{aligned} \tag{8.2}$$

Clearly, $k(s) = 0$ for dipoles and drifts, while the $k(s)$ is k_Q, [Eq. (8.3)] for quadrupoles. The dimension of k is that of inverse length squared and $k(s)$ is periodic, $k(s+C) = k(s)$, by definition. In the literature this is identified as Hill's equation. It is simply a generalization of the equation for a mass on a spring, but one with a variable spring constant. An accelerator has solutions with periodic boundary conditions.

The quadrupole equations shown previously are an example with a constant value of k, while the drift space has a k of zero. Ignoring the spread of the momentum of particles in the beam, dipole magnets are also treated as drift spaces and all particles have the same momentum, $k_d = k_B = 0$. The independent variable is taken to be distance along the reference trajectory, $ds = v\,dt$. The reference orbit is assumed to be in the (x, s) plane. Dipoles are assumed to have fields only in the y direction, so that the central orbit is planar.

In general the solutions will be assumed to be quasiperiodic harmonic oscillator functions. A plausible ansatz is to assume that they have an amplitude and phase which are s-dependent and are of the form

$$x(s) = A\sqrt{\beta(s)}\cos[\psi(s) + \psi_o]. \tag{8.3}$$

The constant A is defined by the beam emittance, which will be defined a bit later. The $\beta(s)$ function is defined by the lattice and is periodic. The phase $\psi(s)$ is not necessarily periodic. The contours of $\sqrt{\beta(s)}$ are the "envelope" of the beam, $|x(s)| < A\sqrt{\beta(s)}$. The assumed form of the solution satisfies the equations of motion, as can be established by direct substitution.

Consider a vector with components α and β in a set of basis vectors defined by the eigenvectors. The eigenvectors are orthogonal and any vector can be expanded in them as a basis set. The transport matrix can then be simply written as a rotation in this basis. After n periods of rotation the phase evolves to $n\mu$. In order to avoid resonant behavior due to the oscillations being reinforced on subsequent revolutions, μ should not be 2π or a multiple thereof.

$$M\begin{pmatrix}\alpha\\ \beta\end{pmatrix} = \begin{pmatrix}\cos\mu & \sin\mu\\ -\sin\mu & \cos\mu\end{pmatrix}\begin{pmatrix}\alpha\\ \beta\end{pmatrix}. \tag{8.4}$$

In Cartesian coordinates rather than the eigenvector basis, the rotational circle becomes an ellipse, somewhat like the shape shown in Figure 8.3. Over many turns the average phase angle advance in the Cartesian basis is still the tune. The transfer matrix can be defined in a conventional manner:

$$\begin{aligned} &M = I\cos\mu + J\sin\mu, \quad I = \begin{pmatrix}1 & 0\\ 0 & 1\end{pmatrix}\\ &J = \begin{pmatrix}\alpha & \beta\\ -\gamma & -\alpha\end{pmatrix}, \quad \beta\gamma - \alpha^2 = 1. \end{aligned} \tag{8.5}$$

The 2×2 matrix M has three independent elements since $|M| = 1$ is a constraint. The determinant of the M matrix remains 1, from which follows the definition of γ. The three independent parameters can be taken to be α, β, and μ. It is unfortunate that the β and γ parameters might be confused with the relativistic terms defined previously, but the conventions for both areas of physics will be respected and will not, one hopes, cause any confusion.

The phase is determined by the s dependence of $\beta(s), \psi(s) = \int ds/\beta(s)$, which can be proved by substituting the assumed solution, Eq. (8.3), into the Hill equation, differentiating twice, collecting terms with the factor $\cos(\psi+\psi_o)$ and $\sin(\psi+\psi_o)$, and requiring both factors to vanish for an arbitrary phase, which yields $d\psi/ds = 1/\beta(s)$ for the first factor. Thus, $\beta(s)$ is not only the "envelope" of the beam but also the local wave number because it also determines the phase advance. A defined quantity which is useful is $\alpha = -(1/2)d\beta/ds$. It follows that β is an extremal when α vanishes. If it is a minimum, then the $\alpha = 0$ location is called a "waist". The maximum is often the limiting aperture location. The phase advance over one period, defined to be the circumference C, is $\mu = \psi(s) - \psi_o = \psi(C)$.

The solution for the system x and x' in terms of the initial vector and the transfer matrix M for the circumference C or a repetitive unit cell follows from Eq. (8.5).

$$\begin{aligned}\begin{pmatrix} x \\ x' \end{pmatrix}_{s_o+C} &= M(s_o+C|s_o)\begin{pmatrix} x_o \\ x'_o \end{pmatrix}_{s_o} \\ &= \begin{pmatrix} \cos\mu+\alpha_o\sin\mu & \beta_o\sin\mu \\ -\gamma_o\sin\mu & \cos\mu-\alpha_o\sin\mu \end{pmatrix}\begin{pmatrix} x_o \\ x'_o \end{pmatrix}_{s_o} \qquad (8.6) \\ \mu = \psi(s)-\psi_o &= \int_{s_o}^{s_o+C} ds/\beta(s). \end{aligned}$$

The tune, μ, is independent of s. For a circumference C consisting of N unit cells, the total tune is $N\mu$ per revolution with a frequency $\upsilon = N\mu/2\pi$ which is also called Q in what follows, but both symbols appear in the literature. In this text Q will be used; $Q = (1/2\pi)\oint ds/\beta(s) = N\mu/2\pi$. Equation (8.6) is derived by appealing to the periodicity of the problem. The meaning is that, for an initial value of x_o and x'_o, after one circumference of distance C, the beam particle is at x and x'.

In general, the "tune" is the total betatron phase advance divided by 2π in one revolution for the two transverse motions, defined to be Q_x and Q_y. A larger tune implies stronger focusing power and a smaller β. Note that the orbit is not re-entrant in one revolution and the tunes are not necessarily integers and should not be to avoid resonances. For example, the Main Ring has about 100 cells each, with a phase advance of about 71°. Therefore, oscillations repeat in about five cells and the total tune is ~19.4. From Eq. (8.6) the

```
>> Q_Tune
  graphic movie for betatron oscillation
  periodicity is not the circumference


Enter the Tune : 5.3
```

Figure 8.1: User input for the movie supplied by "Q_Tune".

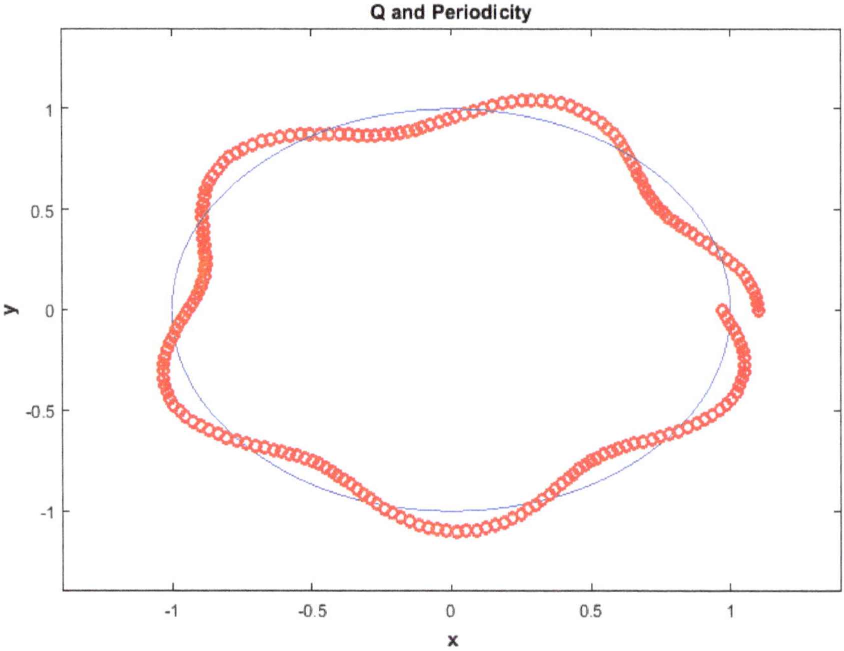

Figure 8.2: The last frame of the movie created by the script "Q_Tune". The orbit does not repeat after one revolution.

initial transverse position and divergence, after a traversal of the full circumference C, do not repeat, which can be made more visual using the "Q_Tune" movie.

A movie of the tune in orbiting the circumference C is provided by running the script "Q_Tune".

The value of the beta function, alpha parameter, and gamma parameter at the cell boundaries are defined by the cell transfer matrix M. If the transport matrix is computed for a unit cell, then the three parameters are completely defined.

$$\beta_o = M(1,2)/\sin\mu, \quad \alpha_o = M(1,1) - \cos\mu/\sin\mu, \quad \gamma_o = 1 + \alpha_o^2/\beta_o.$$

In the symbolic FODO solution shown in "FODO_Ellipse", (Figure 8.1), the trace of the transport matrix is $2 - d^2/f^2 = 2\cos(\mu)$. Since $\cos\mu = 1 - 2\sin^2(\mu/2), \sin(\mu/2) = d/2f$. The FODO stability condition is then that the sine is less than 1 or $f > d/2$. That

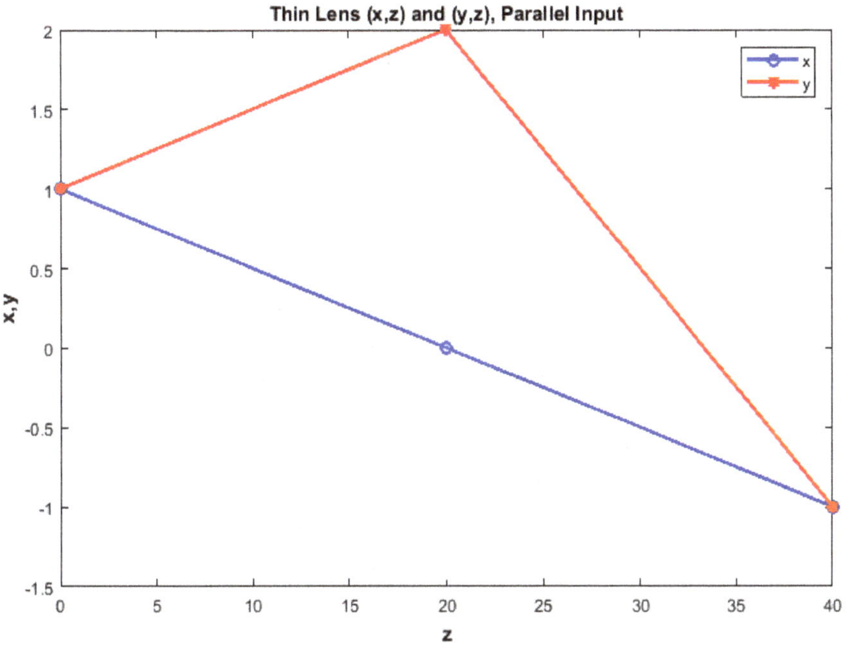

Figure 8.3: Thin lens rays for a parallel beam input on a FODO with $d = 20$ and $f = 10$ which is on the boundary of stability. Note that the initial and final positions have the same magnitude.

condition was previously found in Chapter 7 by asking for net focusing in the FODO. There are other useful trig identities supplied in Appendix D which the reader may wish to consult.

The physical picture of the instability can be explored using the script "FODO_Ellipse". The thin lens FODO with $d = 20$ and $f = 10$ appears in Figure 8.3. The rays at the exit to the FODO are the same as those at the entrance in both x and y, and there is no net focusing. Any weaker focusing will lead to a loss of net overall FODO focusing. Formally, stability requires a phase advance for the cell of $\mu < \pi$.

The β function in the center of the F quadrupole, the FODO boundary, can be found using $M(1,2) = \beta \sin \mu = d(d + 2f)/f$ and $\sin(\mu/2) = d/2f$. The symbolic expression for $M(1,2)$ appears in Figure 8.1. The maximum value of β will occur at the focusing quadrupole, as was seen in the previous doublet study. That is the reason to define the end of the cell as the centerline of the F quadrupole aside from the reason of symmetry. The minimum will similarly occur at the center of the defocusing quadrupole.

$$\begin{aligned}\beta_{\text{max}} &= 2d(1 + \sin\mu/2)/\sin\mu \\ \beta_{\text{min}} &= 2d(1 - \sin\mu/2)/\sin\mu.\end{aligned} \tag{8.7}$$

The dimensions of β are taken to be length, as seen in Eqs. (8.6) and (8.7). The scale for β is $2d$. From Figure 8.1 it can be seen that $M(1,1) = M(2,2)$. This fact means Eq. (8.6), $\alpha = 0$, and the determinant constraint, Eq. (8.5), then implies that $\beta = \beta_{\text{max}}$ and $\gamma = 1/\beta$ at the center of the F quadrupole. The values of the β function depend only on the lattice devices and not on the beam.

The plot of β as a function of the cell phase advance is given in Figure 8.4. The differences are not extreme, which is useful in reducing the maximum, aperture-filling, values. The choice of a

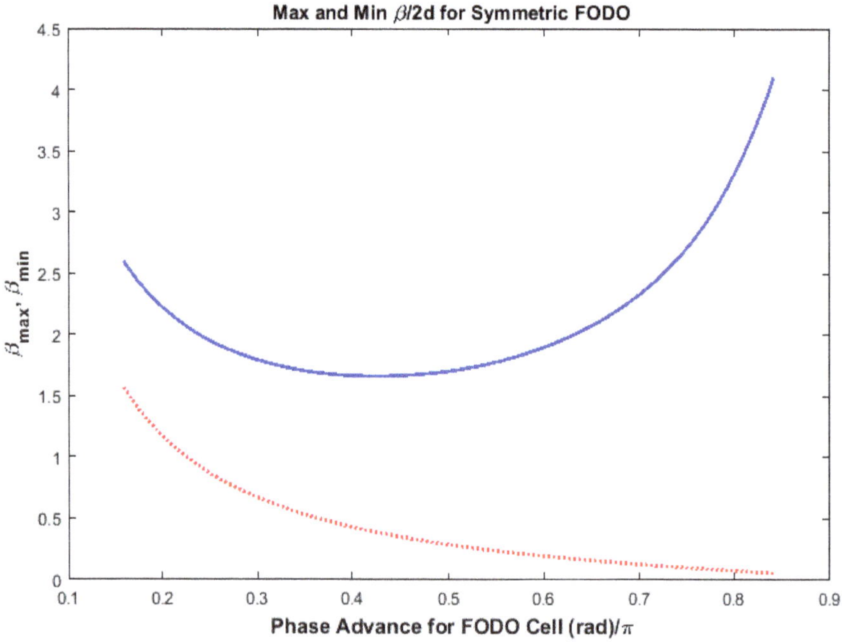

Figure 8.4: Minimum (dotted) and maximum (solid) values of the β parameter in units of $2d$ for a symmetric FODO as a function of the phase advance of the cell.

Table 8.1: Fermilab Main Ring FODO Parameters.

Phase advance (degrees) μ	71
Quadrupole length (m)	2.1
Distance between quadrupoles — d(m)	29.7
$\beta_{\max}$(m)	99.4
$\beta_{\min}$(m)	26.4
Quadrupole focal length (m)	25.6
Quadrupole gradient (T/m)	24.5
For Tevatron quadrupoles (T/m)	77

phase advance of about 90° results in a relatively small value of the excursions of the envelope. The unstable point at $\mu = \pi$ is also evident.

In this text the Fermilab Main Ring is used as the example FODO. Therefore, some properties of the unit cell are presented in Table 8.1 for ease of reference. Other accelerator parameters are shown in Appendix E. The Main Ring FODO was shown previously in Figure 8.5.

It happens that the symmetric FODO is not always the optimal choice. In the more general case of an asymmetric FODO with a distinct f_F and f_D, instead of the two quadrupoles having the same focal length, the trace is calculated symbolically in "FODO_Asymmetric". Then the stability conditions on f_F for $\cos\mu = 1$ and -1 at the stability boundary are computed using the "solve" utility. The printout from this script appears in Figure 8.5.

The stability lines for the asymmetric FODO are shown in Figure 8.6. Any point inside the boundaries will be stable. The diagonal represents the previous solution for a symmetric FODO. The contours of stability are set by $d/f_F = 2$ or $d/f_D = 2$. Using the trace calculation shown in Figure 8.3 the sine of half the phase advance sets the stability limit when it vanishes, $\cos\mu = 1$, or when it is one ($\cos\mu = -1$). In general the phase advance depends on d, f_F, and f_D as

$$\sin^2(\mu/2) = d[d/4f_F f_D - 1/2(1/f_F - 1/f_D)]. \tag{8.8}$$

The symmetric case is recovered when the F and D focal lengths are equal.

```
>> FODO_Asymmetric
  thin lens FODO doublet - symbolic


Transfer matrix for (x,dx/dz)

ans =

[ -(2*d*fD - 2*d*fF - 2*fD*fF + d^2)/(2*fD*fF),                                    (d*(d + 2*fD))/fD]
[   ((d - 2*fF)*(d + 2*fD - 2*fF))/(4*fD*fF^2), -(2*d*fD - 2*d*fF - 2*fD*fF + d^2)/(2*fD*fF)]

Trace = 2cosu
                                    2
  2 d fD - 2 d fF - 2 fD fF + d
- -------------------------------
             fD fF

Stability Lines for fF
d
- + fD
2

d
-
2
```

Figure 8.5: Symbolic solution for stability of an asymmetric FODO with differing focal lengths for the F and D quadrupoles.

There is an invariant ellipse in (x, x') and in (y, y') which defines the stable phase space population at points along the system. A first hint of the ellipse appeared in Chapter 8, when we were exploring the FODO acceptance. The area of the ellipse is invariant but the shape and orientation can vary. The boundary of the ellipse defines the unstable region and is called the dynamic aperture. The invariant area is called the emittance, ε. The ellipse has the form $\gamma x^2 + 2\alpha x x' + \beta (x')^2 = \varepsilon$. This invariant is analogous to the total energy — potential and kinetic — of a simple harmonic oscillator or any other conservative system in classical mechanics.

The acceptance of a system may be larger than the emittance if the physical apertures are larger than the needed apertures, defined by β. As will be seen later, under acceleration of the beam, the emittance shrinks and will fall below the acceptance required at injection at lower energies. The ellipse area is fixed at a value $\pi\varepsilon$. At the maximum beam extent, $\gamma x^2/\varepsilon + \beta(x')^2/\varepsilon = 1$ with axes $a = \sqrt{\beta\varepsilon}$, $b = \sqrt{\varepsilon/\beta}$. The total width at this location is $2a$ in x and $2b$ in x'. An ellipse has area πab, where a and b are the ellipse axes.

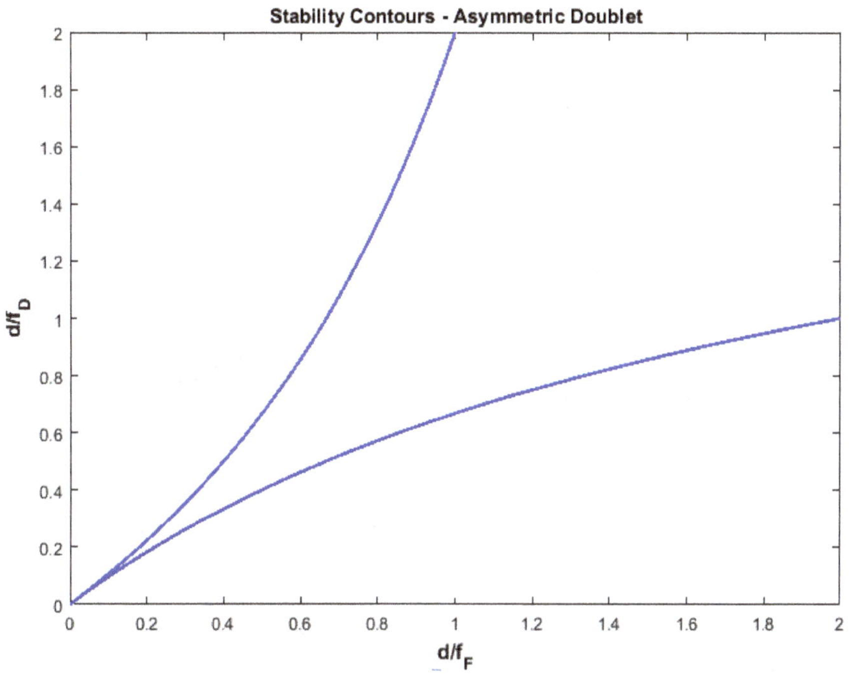

Figure 8.6: Stability boundaries for an asymmetric FODO.

In the absence of dissipation or acceleration, the emittance is a constant of the motion. The units of emittance are m*rad, but more useful units are perhaps mm*mrad. The value for the Fermilab Main Ring is normally defined to be 12π*mm*mrad. For a 1 cm aperture, angles of approximately 3.8 mrad are therefore captured. The approximate precursor to the ellipse appears in Figure 8.3, where the region of phase space with a limiting aperture within a thin lens FODO is shown. As the beam evolves over many rotations and encounters many apertures, it is easy to imagine how the sharp edges seen in Figure 8.3 might evolve into a rounded ellipse shape.

Setting $x = 0$ or $x' = 0$, the intercepts of the ellipse are x_{int} and x'_{int}, as shown in Figure 8.7. The maximum values on the ellipse x_m and x'_m are also easily evaluated. For ellipses oriented in the first or third quadrant, the particles are diverging, while those inhabiting the second and fourth quadrants are converging. A vertical ellipse is

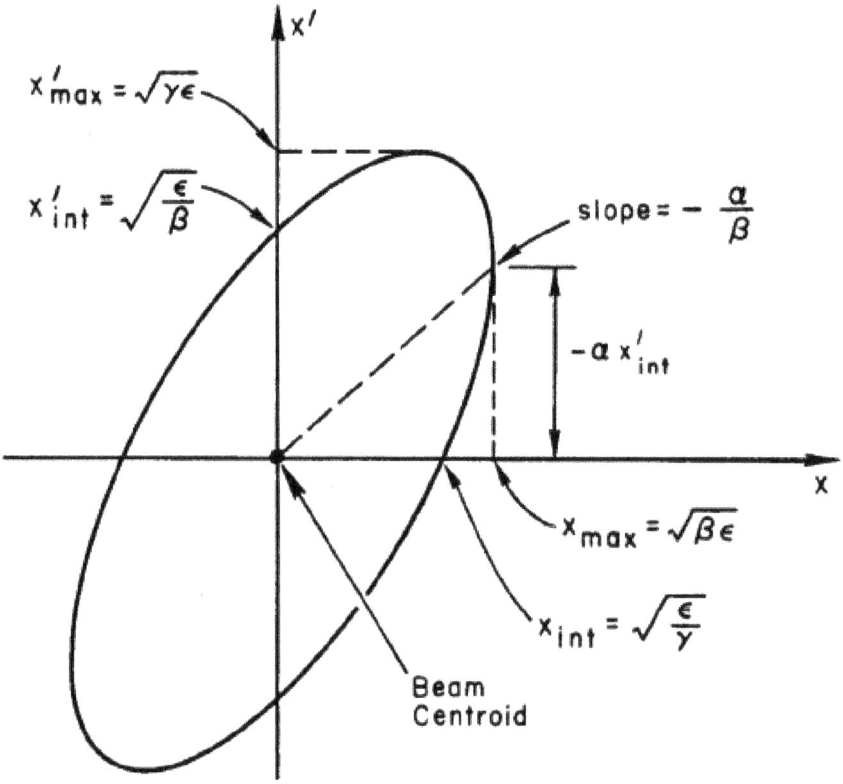

Figure 8.7: The invariant ellipse defined in terms of the transfer matrix parameters, α, β, and γ, and the emittance ε.

near a focus, while a horizontal ellipse is near its maximum spatially.

$$\begin{aligned} x_{\text{int}} &= \sqrt{\varepsilon/\gamma}, \quad x_{\text{max}} = \sqrt{\varepsilon\beta} \sim \sigma_{\text{RMS}} \\ x'_{\text{int}} &= \sqrt{\varepsilon/\beta}, \quad x'_{\text{max}} = \sqrt{\varepsilon\gamma}. \end{aligned} \tag{8.9}$$

The maximum value calculation can be easily done, as illustrated in Figure 8.8, using the symbolic logic tools "diff", "solve", and "subs" to clean things up by imposing the determinant identity in a few command line inputs. The ellipse area is $\pi x_{\text{int}} x'_{\text{max}} = \pi x_{\text{max}} x'_{\text{int}}$. A minimum (or maximum) extent of x occurs when $\alpha = 0$, which is called a "waist" condition. The waist location in a system need

```
>> syms a b g x xp e ee xps xmx
>> e = g*x*x+2*a*x*xp+b*xp*xp;
>> xps=solve(diff(e,xp),xp);
>> e = g*x*x+2*a*x*xps+b*xps*xps;
>> xmx = solve(ee-e==0,x);
>> subs(xmx,- a^2 + b*g,1)

ans =

  b*(ee/b)^(1/2)
 -b*(ee/b)^(1/2)
```

Figure 8.8: Use of symbolic logic to find the maximum value of x attained by the invariant ellipse. The maximum value of x' can be found similarly.

not be the focus, which was instead defined by a requirement on the transfer matrix itself.

The ellipse can be written in a matrix form by defining the elements of a σ matrix and using the definition of the ellipse in terms of the α, β, and γ parameters:

$$\gamma x^2 + 2\alpha x x' + \beta (x')^2 = \varepsilon$$
$$\begin{pmatrix} x & x' \end{pmatrix} \begin{pmatrix} \gamma & \alpha \\ \alpha & \beta \end{pmatrix} \begin{pmatrix} x \\ x' \end{pmatrix} = \varepsilon = \begin{pmatrix} x & x' \end{pmatrix} \sigma \begin{pmatrix} x \\ x' \end{pmatrix}. \tag{8.10}$$

The matrix multiplication results in a scalar, or invariant, quantity. That result serves to establish the invariance of the ellipse. The proof of Eq. (8.10) follows after a few command line operations, as shown in Figure 8.9. Looking at Figures 8.7 and 8.9, it becomes clear that MATLAB removes the tedium of many algebraic calculations.

The dimension of β is length, as is the dimension of the emittance. The dimension of γ is inverse length and α is dimensionless. Typically, the ellipse is not uniformly populated with particles. If the particles are Gaussian-distributed, then the fraction contained in the beam RMS can be found using the MATLAB utility "erf" to evaluate the error function, which yields the fractional area within some upper limit of a Gaussian-distributed quantity. For example, the beam RMS, σ, is often defined, [Eq. (8.9)], to be

```
>> syms x xp a b g e
>> e = [x xp]*[g a ; a b]*[x ; xp];
>> simplify(e)

ans =

g*x^2 + 2*a*x*xp + b*xp^2
```

Figure 8.9: Command line operations to establish the matrix representation of the ellipse.

$\sqrt{\beta\varepsilon}$ and the divergence RMS is $\sqrt{\gamma\varepsilon}$. The fraction within $n\sigma$ is just erf(n*0.707). The results for $n = 1$ and 2 are 68% and 95%, respectively.

The machine lattice has defined the ellipse so far with an invariant emittance. A beam ellipse can also be defined by quoting the averages over the particle distributions within the beam. There are clearly correlations between position and angle in general, and therefore the beam is defined by a 2×2 matrix with elements $\sigma_{11} = <x^2>, \sigma_{12} = \sigma_{21} = <xx'>, \sigma_{22} = <(x')^2>$, where the symbol $<>$ implies an average over the particles in the beam. The square of the beam emittance is given by the determinant of the matrix, which takes the $x - x'$ correlations into account.

For a constant aperture $a, \varepsilon = a^2/\beta$. The ellipse boundary is defined by the apertures of the system in question when a beam definition is used, not a lattice definition. The population of the ellipse area can be visualized as a set of nested ellipses of area up to the one defined by the limiting aperture or the emittance itself.

An alternative formulation of the ellipse in terms of the elements of a second σ matrix follows from expressing γ, β, and α in terms of σ_{11}, σ_{22}, and σ_{21} and normalizing to ε. The results are shown in Figure 8.10. It is simple to identify σ_{11} as $\beta\varepsilon$ and σ_{22} as $\gamma\varepsilon$ in this alternative formulation of the matrix. This matrix for the beam ellipse transforms as $\sigma' = M\sigma M^T$, where M is the transport matrix. In this text the first definition of the ellipse in terms of α, β, and γ will be employed exclusively. However, it is useful to be aware of

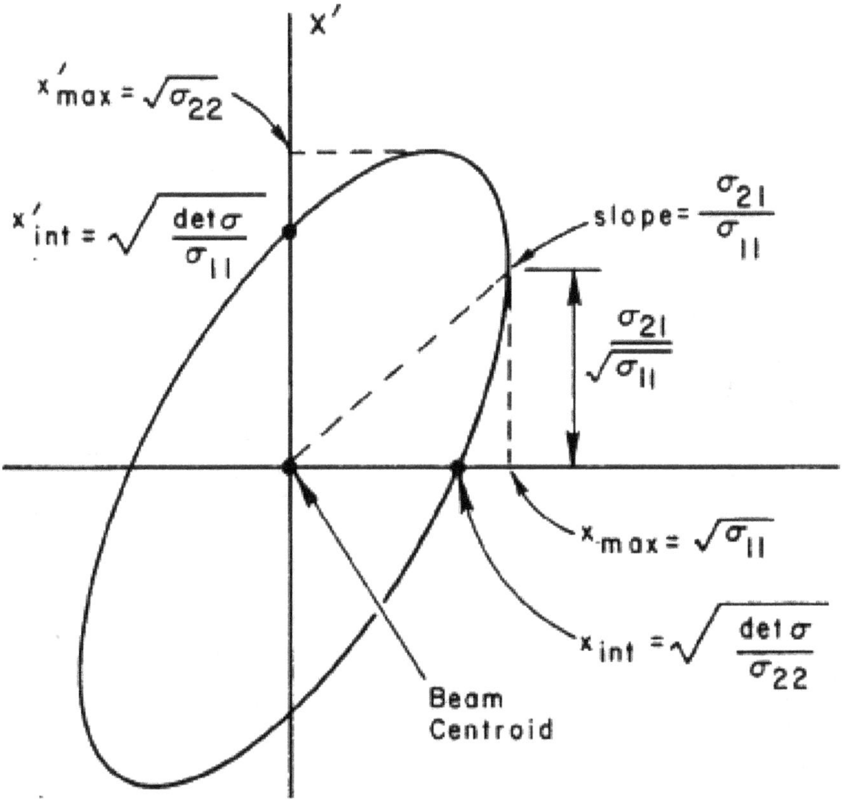

Figure 8.10: The invariant ellipse defined in terms of the matrix elements defining the ellipse in an alternative formulation of the σ matrix using the beam distributions for position and angles themselves.

alternative formalisms so as to avoid confusions and to make the distinction between a lattice-defined ellipse area and a beam-defined ellipse area.

The ellipse and the three lattice parameters have been defined on the boundaries of the unit cell by appealing to the periodicity of the solutions to the equations of motion. It is of interest to see how the ellipse appears at any intermediate positions.

If the ellipse passes through a transport matrix M, the transformed ellipse follows from standard matrix transformations, and is $\sigma' = (M^{-1})^T \sigma M^{-1}$ in this special case with $\det(M) = 1$.

```
>> Ellipse_Propagate
Transformed beta After Transfer Matrix M
                         2        2                    2                      2         2     2
m21 m22 (alf m11   |m12|   - bet m11 m12 |m11|   - gam m11 m12 |m12|   + alf m12   |m11| )
-------------------------------------------------------------------------------------------
                              2      2               2      2
         (m11 m22 |m12|   |m21|   - m12 m21 |m11|   |m22| ) (m11 m22 - m12 m21)
```

Figure 8.11: Transformed values of the β ellipse parameter after traversal of a transfer matrix m. After also finding γ, the value of α follows from the constraint that the determinant of m is 1.

The inverse and transpose operations are used on the transfer matrix M. This operation is carried out explicitly in the script "Ellipse_Propagate".

First, the transformation is done symbolically using the utilities "inv" to find the inverse matrix and the transpose operation ('). The transformed matrix, examining Eq. (8.10), has a (1,1) component which is the new γ paramcter, and the $(2,2)$ component is the new β parameter. The resulting expressions are cleaned up using "simplify" and "pretty", and appear in Figure 8.11.

However, the general case is not very illuminating. Special cases of the transformation of the α, β, and γ parameters for the traversal of a thin lens quadrupole and a drift space are shown symbolically in Figure 8.12. The explicit evolution of β in a drift space will be referred to later, in a discussion of "low beta" insertions. It is notable that the β function evolves quadratically with distance in a drift space. Plotting $\sqrt{\beta}$ makes the dependence linear, which is a reason for deciding to choose to plot this variable.

In the literature there is an alternative definition of the ellipse matrix which results in a second definition of the ellipse transformation. However, the final results are the same, so no confusion should arise. The two transformations are

$$\begin{aligned} \sigma &= \begin{pmatrix} \gamma & \alpha \\ \alpha & \beta \end{pmatrix}, \quad \sigma \to (M^{-1})^T \sigma M^{-1} \\ \sigma &= \begin{pmatrix} \alpha & \beta \\ -\gamma & -\alpha \end{pmatrix}, \quad \sigma \to M \sigma M^{-1}. \end{aligned} \tag{8.11}$$

```
Special Case - Quadrupole, New alpha, beta,  gamma
          bet
alf + -------
        conj(f)

bet

                                              2
                           2    alf |f|
bet + alf f + gam |f|   + --------
                                      f
------------------------------------
                        2
                    |f|

Special Case - Drift, New alpha, beta gamma
alf - d gam

          2
gam |d|   + bet - alf d - alf conj(d)

gam
```

Figure 8.12: Ellipse parameter transformations for the traversal of a thin quadrupole, f, and a drift space, d. The quadrupole does not alter β, while the drift does not change γ.

In the equation, the notation T indicates the transpose of the matrix. In the scripts the transpose is computed using the MATLAB matrix utility " ' ". It should be kept in mind that the M shown here is the transfer matrix to a new location and not the matrix shown in Equation (8.6) for a rotation by the full circumference or repeat length.

In order to better visualize the general ellipse transformations, the script "Ellipse_Propagate" asks the user to choose an initial ellipse by specifying β and γ. That ellipse is plotted and then a choice of a drift space, an F quadrupole or a D quadrupole, is made and the ellipse is transformed through that element. An example of

```
>> Ellipse_Propagate
beta ~ xmax: 2
gamma ~ dx/ds max: 4
x max 1.41421 and dx/dz max 2
new gamma 4 and new beta 7.41699
x max 2.72342 and dx/ds max 2
beta ~ xmax: 2
gamma ~ dx/ds max: 4
x max 1.41421 and dx/dz max 2
new gamma 0.708497 and new beta 2
x max 1.41421 and dx/ds max 0.841723
beta ~ xmax: 2
gamma ~ dx/ds max: 4
x max 1.41421 and dx/dz max 2
new gamma 11.2915 and new beta 2
x max 1.41421 and dx/ds max 3.36028
```

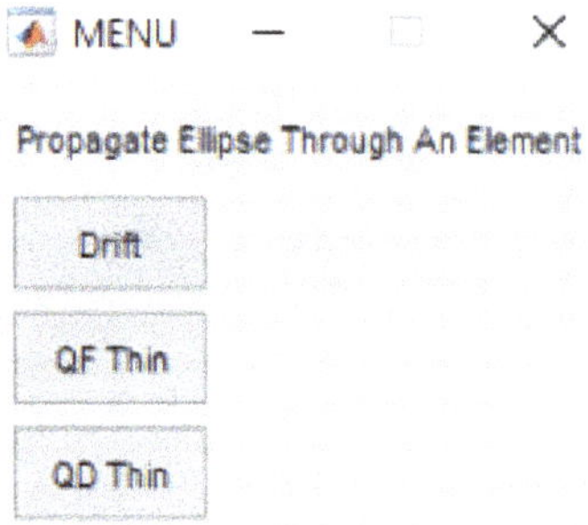

Figure 8.13: An example of the dialogue for "Ellipse_Propagate". The initial and transformed β and γ parameters are shown along with the maximum values of x and x' on the ellipse boundary. The "menu" choices are also displayed.

the dialogue appears in Figure 8.13, while the results are shown in Figures 8.14, 8.15, and 8.16.

The changes in shape are usefully seen to be an elongation of x for a drift space at fixed ellipse maximum x', a reduction in x' for a focusing quadrupole and an extension of x' for a defocusing quadrupole, both at fixed x, as expected. The ellipse area remains constant by construction. Also shown are the transformations, of two fixed points on the ellipse, x_{int} and x'_{int}. Their movement illustrates that individual points move on the ellipse as the trajectory traverses the beam element. In the script α is derived from β and γ using the quadratic relation. There is a sign ambiguity which can be resolved using the fact that $\alpha = -(d\beta/ds)/2$ to associate negative α with beam divergence (Figure 8.7) and positive α with a decreasing value of β, or focusing. The ellipses rotate clockwise as the value of s increases. It is recalled that α is zero at the Q_F and Q_D center, where the β function is an extremal and turns over.

There is an alternative to propagation of rays from an initial point to some arbitrary point in the lattice. The resulting transport matrix is not formulated as $M(1,1), M(1,2), M(2,1)$, and $M(2,2)$

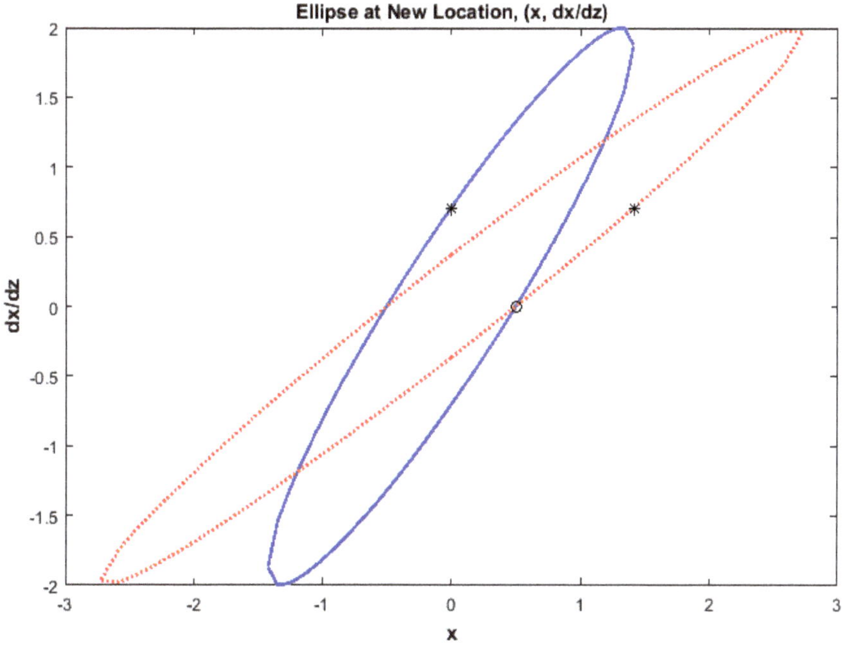

Figure 8.14: Input ellipse (solid) and ellipse after a drift space (dotted). For a drift space the maximum x increases while the maximum x' remains constant. The o points are the initial x_{int} and x'_{int} points of the ellipse intersections with the x and x' axes, while the * points are their transformed locations. This is also true for the two figures that follow.

but using the equivalent variables, α, β, and the phase advance from the initial point (labeled with subscript o), to the final point s with phase advance ψ. The expression is not very transparent, but is simplified if the initial and final points have α of zero, when $\beta = \beta_{\max}$ or $\beta_{\min}$ and $\gamma = 1/\beta$. In that case the rays transform as

$$\begin{bmatrix} x \\ x' \end{bmatrix} = M(s|s_o)\begin{bmatrix} x_o \\ x'_o \end{bmatrix} = \begin{bmatrix} \sqrt{\beta/\beta_o}\cos\psi & \sqrt{\beta\beta_o}\sin\psi \\ -\sin\psi/\sqrt{\beta\beta_o} & \sqrt{\beta_o/\beta}\cos\psi \end{bmatrix}\begin{bmatrix} x_o \\ x'_o \end{bmatrix}. \tag{8.12}$$

It is clear that if the point β is the initial point, then $\psi(C) = \mu$ and Eq. (8.6) is recovered in the case where α and the initial α values are zero. The proof in the general case is postponed to the later discussion of errors in quadrupoles and dipoles. In this text the

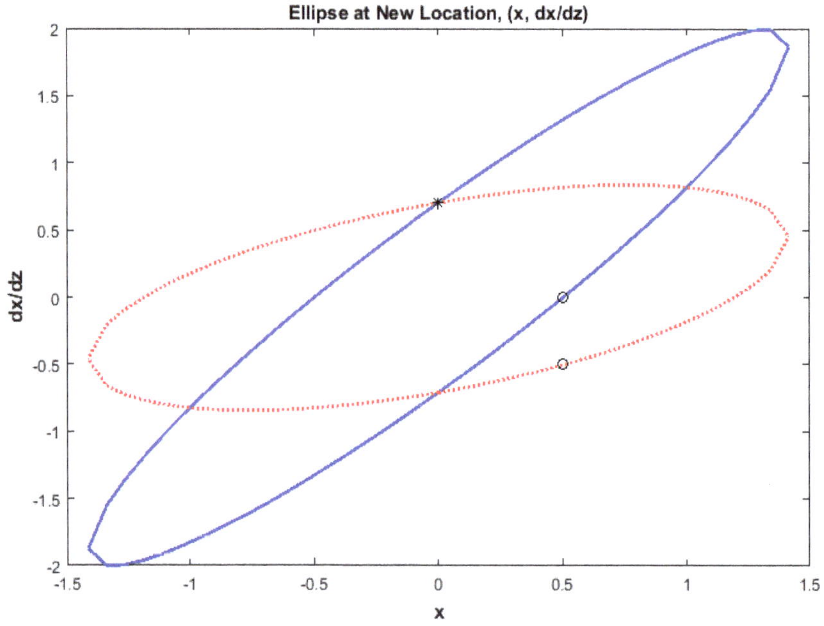

Figure 8.15: Input ellipse (solid) and ellipse after an F quadrupole (dotted). The maximum x stays constant (thin lens), while the maximum x' decreases — focusing.

direct use of the transport matrix will normally be carried out to track rays. Given the one-to-one equivalence of the transport matrix and the ellipse parameters, it is perhaps simplest to use a single formulation of the math. However, both formulations appear in the literature, which makes it incumbent on the user to develop a facility for both. Indeed, it may happen that the transport matrix based on magnets is not known well, but the α and β parameters are measured experimentally by beam monitors that detect the envelope and the change of the envelope with s.

Using Eq. (8.6) to track through an MR FODO with $\alpha = 0, \beta = 100\,\mathrm{m}$, and $\mu = 71^o$, a view of motion of a particle on the ellipse boundary is provided in the script "MR_FODO_Track". The phase is incremented by μ for every FODO traversal, with an expected Q value of 5.09, so that only five FODOs are tracked. A "movie" of six frames is made by the script. The resulting points in (x, x') space for an initial point $(10, 0)$ appears in Figure 8.17.

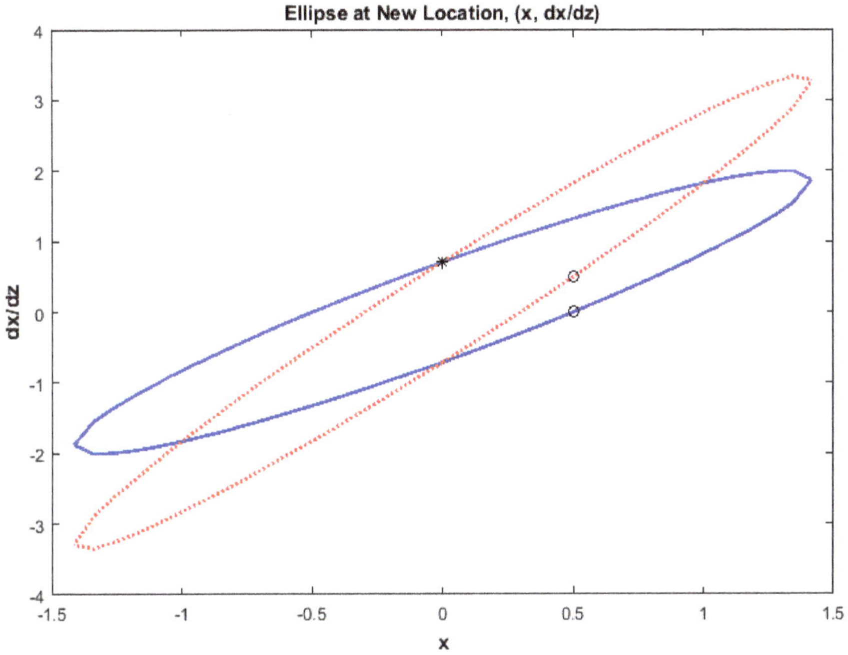

Figure 8.16: Input ellipse (solid) and ellipse after a D quadrupole (dotted). The maximum x stays constant, while the maximum x' increases — defocusing.

The tracking of the β function through an array is a very common practice and provides visualization of both a single particle path and an ensemble of particles which defines an "envelope" of the beam. The envelope repeats in a FODO, but a given ray does not. A specific ray repeats in a number of cells that depends on the overall tune. For the script "FNAL_Main_Ring_FODO" the repetition length is about $2\pi/\mu$, or about five cells ($\mu \sim 71^o$). A plot of the envelope, approximated by linear behavior in between thin lens quadrupoles, appears in Figure 8.18. The dotted line is the ray trace of a single particle, which indeed repeats after about five unit cells. The repeat distance is specific to the MR FODO. The straight line behavior of β in a drift space is not strictly correct, (Figure 8.12), but serves as a first approximation.

If the user chooses to generate more rays, a picture builds up. The envelope of the β function is the limiting aperture and the rays fill in the region between those aperture limits, as seen in Figure 8.19,

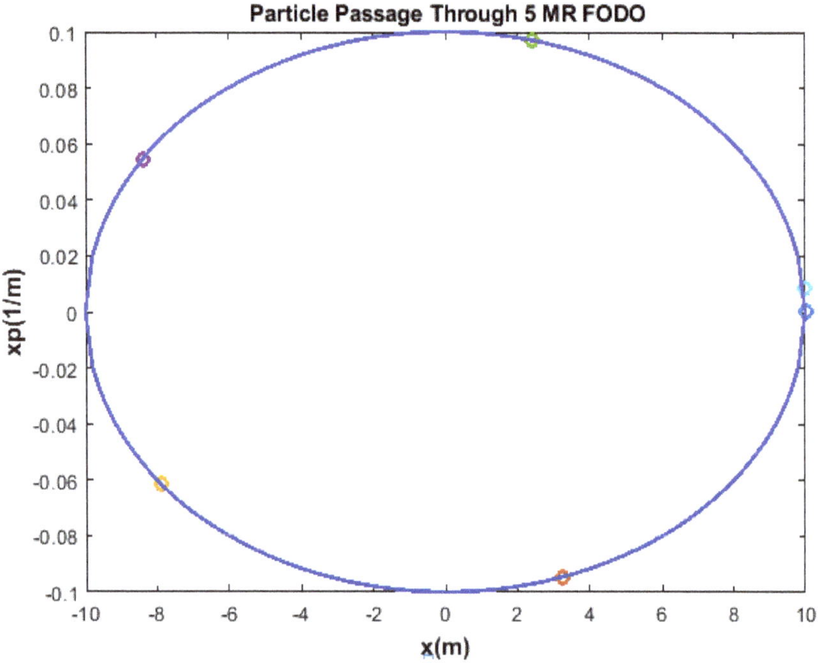

Figure 8.17: Propagation of (x, x') through five MR FODOs. The point for a phase of 5μ almost repeats with the initial point, which has $x' = 0$.

which has 20 rays generated within the MR invariant ellipse. There are small excursions outside the envelope but they are simply the result of the approximations which have been used.

It is also true that when rays outside the acceptance of the FODO are injected, in this case twice the x' range of the acceptance, the aperture is exceeded, as seen in Figure 8.20, which has 10 rays plotted. This is a simple example of behavior when rays are injected outside the phase space acceptance of the lattice, which then increases the beam emittance.

The FODO beta function has been plotted as a simple straight line. That description can be slightly altered by using the actual ellipse transformations in the FODO. For concreteness, the Fermilab Main Ring FODO was chosen as an example. The plotting of β began at the centerline of the F quadrupole, where $\alpha = 0$ and

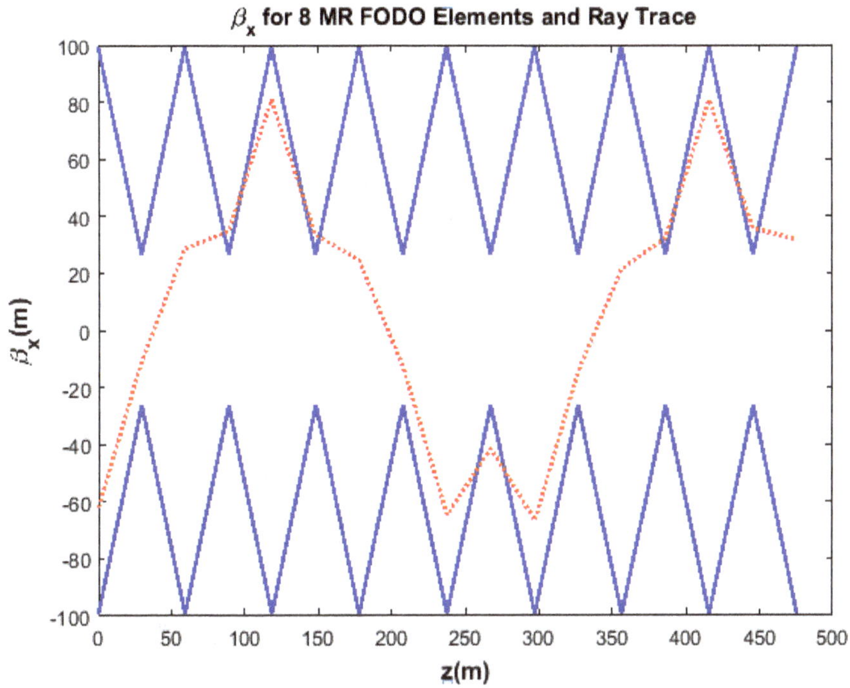

Figure 8.18: Betatron envelope function (solid line) over eight Main Ring unit cells. The dotted line is the ray trace of a single particle within the acceptance of the Main Ring.

$\beta = \beta_{\text{max}}$, since that location has the maximum x extent of the beam. The value for $\gamma = 1/\beta_{\text{max}}$ follows from the $\det(M) = 1$ constraint. The evaluation of β through the FODO is made in the script "Straight_Beta", and the result is shown in Figure 8.21. Clearly, the simple straight line approach gives a reasonable approximation except for details of the aperture needed in the quadrupoles themselves and details of how the beta function evolves in a drift space (Figure 8.12). The cusp behavior in the quadrupoles is softened by accurately tracking the orbit.

The ellipse contours are an alternative to the beta function plots. Switching back to the ellipse representation of the system, a schematic for emittance growth can be studied, as shown in

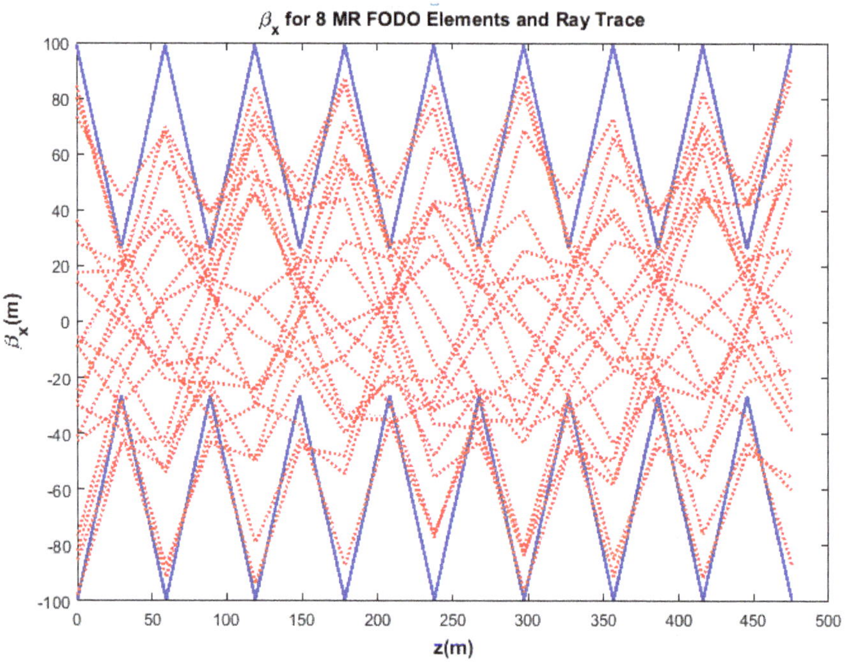

Figure 8.19: Betatron aperture (solid line) and the trajectory of 10 rays (dotted lines) within the acceptance of the MR FODO.

Figure 8.22, complementary to Figure 8.20. In general, the emittance of the system grows with an ellipse mismatch and nonlinear effects are important. Matching the system optics is especially important when injecting or extracting a beam, or when inserting nonperiodic optical elements. A specific case is the insertion of "long straight sections" into a lattice, as will be discussed later.

A simpler example, tracking through 40 MR FODO "turns" using Eq. (8.6), where the final phase is $N\mu$, with $\mu = 71^\circ$ and with $N = 40$ being calculated using the script "Ellipse_Mismatch". The ellipse is taken at the boundary of the MR FODO at the maximum position. A particle with $x = 0$ and dx/ds equal to twice the value allowed with the MR ellipse is injected and tracked for 40 unit cells. The resulting locations of the particle are shown in Figure 8.23. The particle traces

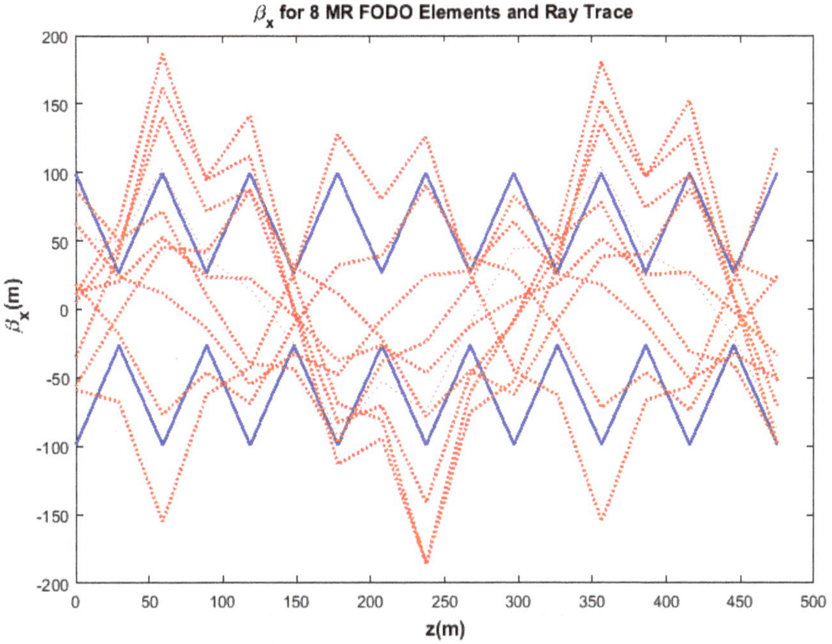

Figure 8.20: Betatron aperture for the Main Ring FODO (solid line) and the trajectory of 10 rays, which can extend to twice the initially accepted x' values (dotted lines).

out the same-shaped ellipse with an expanded emittance. Clearly, any mismatch of the injected beam and the machine lattice implies emittance growth.

The Fermilab Main Ring FODO is again taken as an example and the ellipses within the FODO are plotted at 13 locations in the FODO using the script "MR_FODO_Ellipse". The dialogue appears in Figure 8.24. The drift spaces are cut into four slices to give an indication of how the tracking through a drift space changes the ellipse. A movie of the evolution of the ellipse is provided. (These ellipse plots are used in the cover of this book.)

The user is asked for a choice of location and the appropriate ellipse is plotted. At the end of the users' choices of locations a "movie" of the ellipses going through the FODO is generated.

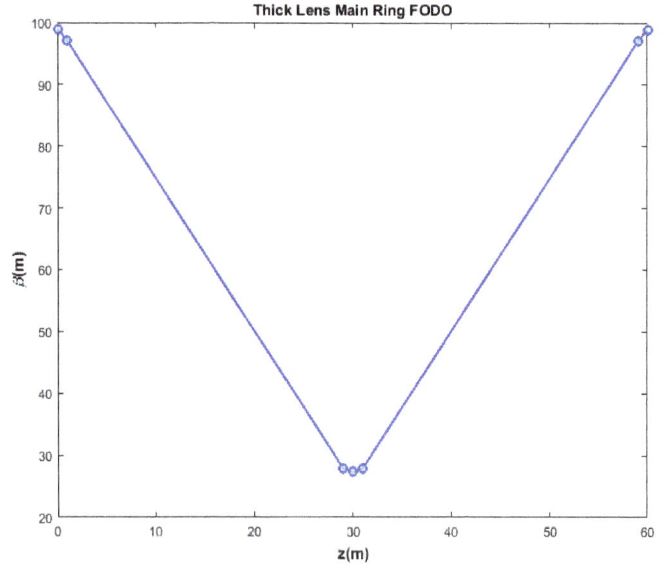

Figure 8.21: Plot of β for the Fermilab MR FODO, with emphasis on the quadrupole values. The straight line approximation in the draft spaces is assumed. The quadrupoles occupy a small fraction of the FODO real estate and the dipoles are taken to be drift spaces. The exact drift space behavior is displayed in Figure 8.12.

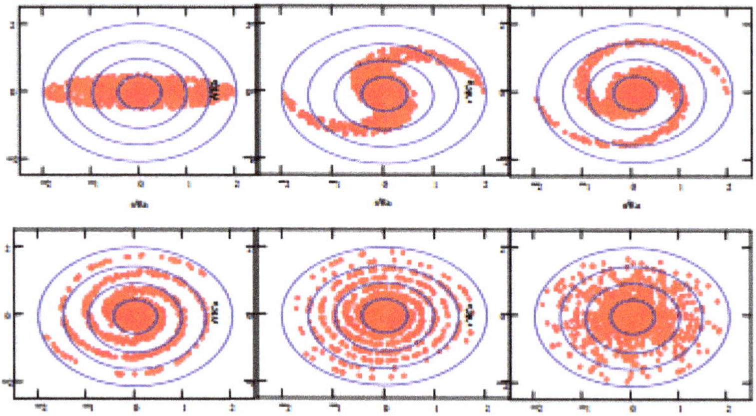

Figure 8.22: Representation of an acceptance ellipse (blue lines) and the injection of a phase space which does not match (red points). The evolution of the injected beam is tracked over many revolutions. Nonlinear terms in the lattice cause the mismatched beam to "dilute" in phase space and the emittance will grow.

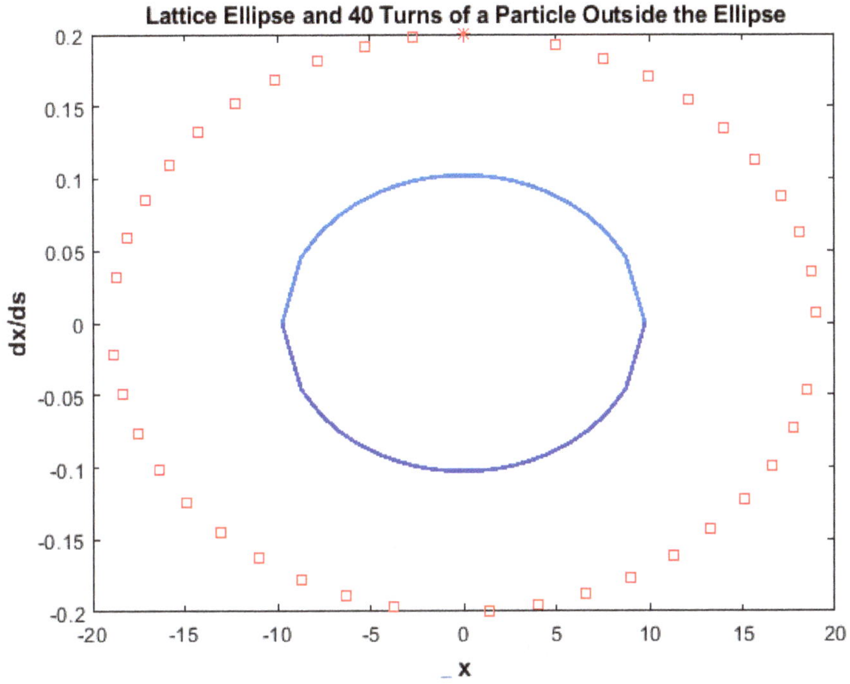

Figure 8.23: The MR FODO ellipse at the FODO quadrupole boundary (solid blue line). The injected particle is shown as a red *. The locations of that particle in the 40 subsequent lattice traversals appear as red squares. A "movie" of the particle position is provided in the script.

The overlapping of the frames of the movie is shown in Figure 8.25. The user can easily edit the script to delete the "hold on" command, which will then simply show sequential "frames" of the movie. This movie summarizes the behavior of the lattice ellipse in the basic FODO unit cell. It bears watching repeatedly. Note that a given particle stays on the ellipse boundary, but goes to a different point on that boundary after each traversal of C (Figure 8.17).

The discussion of periodic structures has so far concerned a boring succession of unit cells composed mostly of dipoles to bend the beam in a circle and thin quadrupoles to economically contain the beam in a vacuum pipe. The interesting items are still to come, where the experiments get done. But first a look at the off-momentum

```
>> MR_FODO_Ellipse
  look at the ellipses in the MR FODO
  step through FODO, drifts in 4 slices


Ellipses in the MR FODO
1=QF/2,2=QF,4=d/2,5=d,7=QD/2,8=QD,10=d/2,12=d,13=QF/2
```

Figure 8.24: Output text of the script "MR_FODO_Ellipse", indicating the points sampled in the FODO.

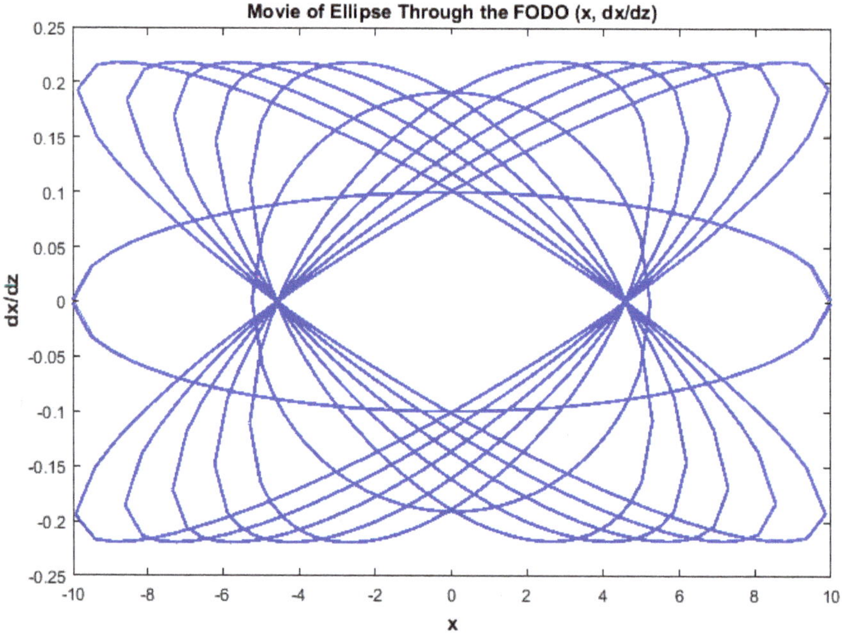

Figure 8.25: Ellipses in the symmetric FODO at 13 locations. The first plot is in the center of the entrance F quadrupole, where x has the maximum extent. The movie follows until the ellipse is in the center of the D quadrupole, where x has the minimal extent, and then follows back to the center of the exit F quadrupole.

particles in the beam is made, since this "complication" has been avoided so far. Specifically, a practical accelerator system must have a finite acceptance not only in the transverse position and angle but also in accepting a finite range of particle momenta.

Chapter 9

Dispersion

> The day when the scientist, no matter how devoted, may make significant progress alone and without material help is past. This fact is most self-evident in our work. Instead of an attic with a few test tubes, bits of wire and odds and ends, the attack on the atomic nucleus has required the development and construction of great instruments on an engineering scale.
>
> — **Ernest O. Lawrence**

The beam has been considered to be monochromatic up until now. However, all actual beams have a finite momentum spread and that needs to be taken into account. The momentum spread leads to a spread in the dipole bend angle, which means that off-momentum particles have a different central orbit than the reference orbit.

A picture of the analogue of this behavior in classical optics appears in Figure 9.1, where the action of a prism on incident white light is displayed.

The dipole trajectory for an off-momentum particle is with respect to the central orbit and depends directly on $(dp/p) = \delta$. The bend angle for small bends is $qeBL/p$, as was derived in the previous chapter on dipoles. The change in the bend angle is $d\phi = -(qeBL/p^2)dp = -\phi_B(dp/p)$. Taking the change in the bend to occur at the center of the dipole, a thin element approximation, there is a shift in the exit position by $-\phi_B\delta(L/2) = -\rho\delta\phi_B^2/2$.

The assumed dipole shape is a "sector dipole", as shown in Figure 9.2, where the magnet is bent into a sector of a circle so that the reference orbit exits the magnet perpendicular to the magnet face. In any case fringe effects will be largely ignored in this text.

Figure 9.1: The result of passing white light through a prism is to spread out the incident beam since the prism bends light of the different wavelengths contained in the white light by different angles, in analogy with what happens in a dipole magnet.

They are covered in many of the references given at the end of the text. These two terms for the shift in the exit angle and position are the small angle versions of the dipole transfer matrix shown in Eq. (9.1).

It is assumed that the reference orbit is planer and is in the (x, s) plane. Dipoles are oriented so that there is no dispersion in the (y, s) plane. The transfer matrix formalism is therefore extended only to three dimensions in the (x, s) plane. Also, any dispersive effects due to the quadrupoles are ignored since the dipoles create much larger dispersion in general.

$$M_B X = \begin{pmatrix} \cos\phi_B & \rho\sin\phi_B & \rho\,(1-\cos\phi_B) \\ -(1/\rho)\sin\phi_B & \cos\phi_B & \sin\phi_B \\ 0 & 0 & 1 \end{pmatrix} \begin{pmatrix} x \\ x' \\ \delta \end{pmatrix}. \quad (9.1)$$

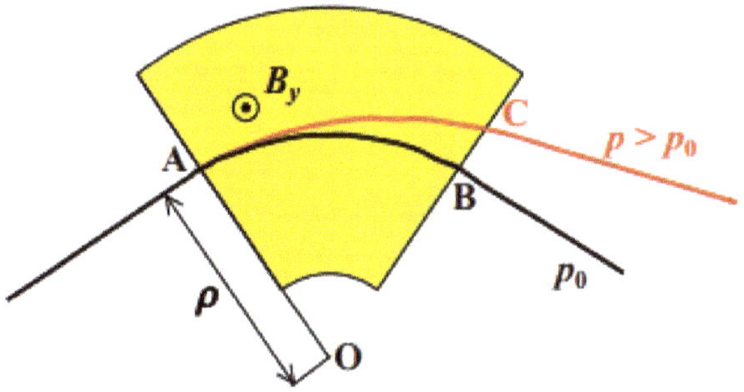

Figure 9.2: Off-momentum orbit in an ideal sector dipole. There are momentum-dependent changes in both the exit angle and the exit position.

The matrix description is now expanded to become a 3×3 matrix in order to accommodate off-momentum effects. The small bend angle approximation for the sector dipole transfer matrix ignoring fringe fields with L the dipole length in $s, \phi_B = L/\rho$, is

$$M_B \sim \begin{pmatrix} 1 & L & \phi_B L/2 \\ 0 & 1 & \phi_B \\ 0 & 0 & 1 \end{pmatrix}. \tag{9.2}$$

Using this simplified approximate dipole transfer matrix, a symbolic solution of the FODO transfer matrix is created in the script "FNAL_MR_FODO_Dispersion". The printout of that script appears in Figure 9.3. The full transfer matrix is quite cumbersome, so only the $M(1,3)$ and $M(2,3)$ terms are printed. The scale of the off-momentum position is set by $d\phi_B$. In this script, in the future, L is used only to define a quadrupole or sextupole length, while d defines both a drift and dipole length since often the dipoles have been treated as approximate drift spaces.

This rather unwieldly matrix has an $M(1,3)$ element which shows the contribution to x at the end of the FODO due to dispersion. The $M(1,3)$ element is proportional to $d\phi_B$. It has been assumed that the entire previous drift space of length d has been replaced by dipoles of the same length but which are still labeled d. From Figure 9.3 the change in x due to dispersion alone selects out the change in the reference orbit, assumed to be $(0,0)$, owing to the momentum

```
>> FNAL_MR_FODO_Dispersion
  FNAL MR FODO - Matrices and ray traces with dispersion


 Thin Lens Doublet with Dipoles Symbolic
MT Transfer Matrix, MT(1,3), MT(2,3)
d phiB (d + 4 f)
----------------
       2 f

         2              2
  phiB (d  + 2 d f - 8 f )
- ------------------------
              2
           4 f
```

Figure 9.3: Printout of the script "FNAL_MR_FODO_Dispersion", which displays the full transfer matrix, of which only the $M(1,3)$ and $M(2,3)$ elements are shown here.

spread, and is

$$\delta x = (\phi_B d)(d/2f + 2)\delta. \tag{9.3}$$

The initial position and angle are chosen at the extremes of the MR FODO ellipse. With the MR emittance of 3.8×10^{-5} m*rad, and with $\beta_{\max} = 100$ m, $x_{\max}$ is 0.058 m and $x'_{\max}$ is 0.56 mrad at the Q_F centerline (Figure 9.7). The δx values for the FNAL MR FODO are shown in Figure 9.4. Excursions of several cm occur for a few-percent momentum acceptance in the FODO. The scale for displacement is $2d\phi_B\delta$, which for the Main Ring FODO is about 7.6 cm for a dp/p of 1%.

This introduction to FODO dispersion establishes the relevant order-of-magnitude scales. Now proceeding somewhat more formally, dispersion induces a driving term into the Hill equation since it changes the reference orbit, $d^2x/d^2s + k(s)x = (dp/p)/\rho$. The dispersion is defined to be $\eta(s)$. It is the change in the reference orbit, dx, with fractional momentum deviation from the mean, $\delta = dp/p, \eta = dx/d\delta$. There is an induced additional term in the

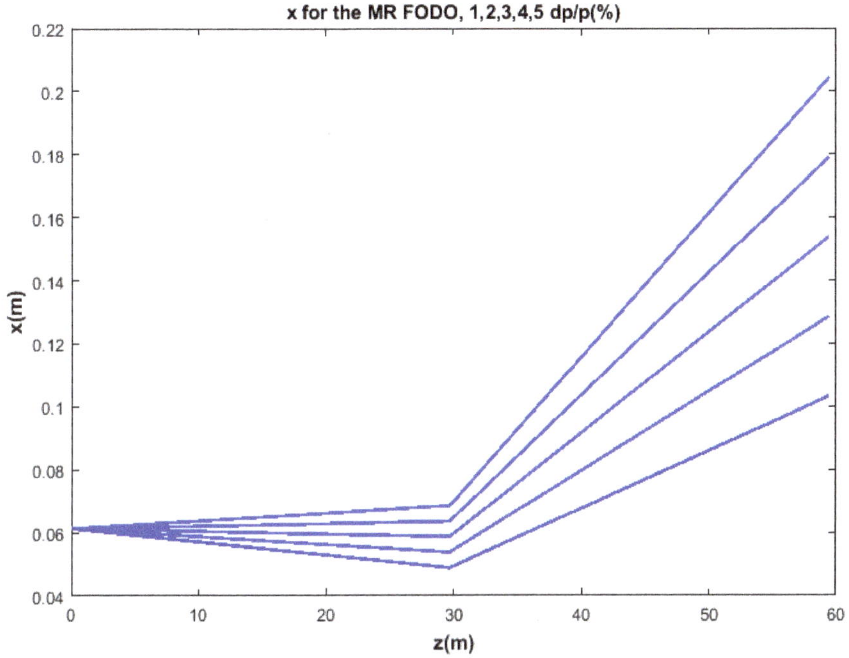

Figure 9.4: Ray traces in the Fermilab Main Ring FODO for rays with initial x and dx/ds values on the extremes of the envelope ellipse and, in addition, a few percent off-momentum.

resulting solution for the inhomogenous Hill equation. The function $\eta(s)$ is periodic and has the dimension of length.

The dispersion functions are the $M(1,3)$ and $M(2,3)$ elements of the transfer matrix for (x,s) motion. They are defined to be $D(s)$ for $x(s)$, and $D'(s)$ for $x'(s)$. The matrix formalism is extended to three dimensions. The off-momentum beam particles cause a displacement in addition to the betatron oscillations of

$$\begin{pmatrix} x \\ x' \\ \delta \end{pmatrix} = \begin{pmatrix} 2\times 2 & & D \\ & & D' \\ 0 & 0 & 1 \end{pmatrix} \begin{pmatrix} x_o \\ x'_o \\ \delta \end{pmatrix}, \quad x=\eta, \quad x'=\eta', \quad \delta = 1. \tag{9.4}$$

The formalism for the on-momentum beam component is represented by the "2 × 2" submatrix and is unchanged, since the chromatic effects of the quadrupoles are small and ignored. The previous on-momentum results are not therefore affected and previous lattice results are employed.

Using the symbolic results for the FODO with magnets, Figure 9.3, worked out as MT in the script:

$$\begin{aligned} M(1,3) &= D = 2d\phi_B(1 + d/4f) = 2d\phi_B[1 + \sin(\mu/2)/2] \\ M(2,3) &= D' = \phi_B[2 - (d/2f) - (d/2f)^2] \\ &= \phi_B[2 - \sin(\mu/2) - \sin^2(\mu/2)]. \end{aligned} \tag{9.5}$$

The previous FODO betatron results, Eq. (9.6), are unchanged and are used to express the D and D' functions as functions of the FODO phase advance:

$$\begin{aligned} \sin(\mu/2) &= d/2f \\ \beta &= 2d[1 \pm \sin(\mu/2)]/\sin(\mu). \end{aligned} \tag{9.6}$$

The FODO transfer matrix has $\alpha = 0$ at the boundaries — the Q_F centerline. The equation to solve for imposition of the periodicity requirement for η, Eq. (9.4), then simplifies to $\eta = \eta\cos\mu + \eta'\beta\sin\mu + D = 0$, but η' is zero at the Q_F centerline, and so $\eta = -D/(1-\cos\mu)$, which, using the transfer matrix element $M(1,3)$ in Figure 9.3 to express D, simplifies to

$$\begin{aligned} \eta_{\max} &= (d\phi_B)[1 + (1/2)\sin(\mu/2)]/\sin^2(\mu/2) \\ \eta_{\min} &= (d\phi_B)[1 - (1/2)\sin(\mu/2)]/\sin^2(\mu/2). \end{aligned} \tag{9.7}$$

A plot of the extreme values of η is made as a function of the FODO phase in the script "FODO_beta_dp", which also plots the β and η functions numerically for the MR FODO. The symbolic plot is displayed in Figure 9.5, while the numerical values of β and η times 50 appropriate to the MR FODO are shown in Figure 9.6 as a function of z in the FODO.

For the FNAL MR FODO phase advance, about 71°, the dispersion minimum is 2.06 m and the maximum is 3.75 m, or 3.75 cm per 1% momentum spread δ or dp/p. This maximum excursion occurs in the F quadrupole centerline. For comparison, the dispersion

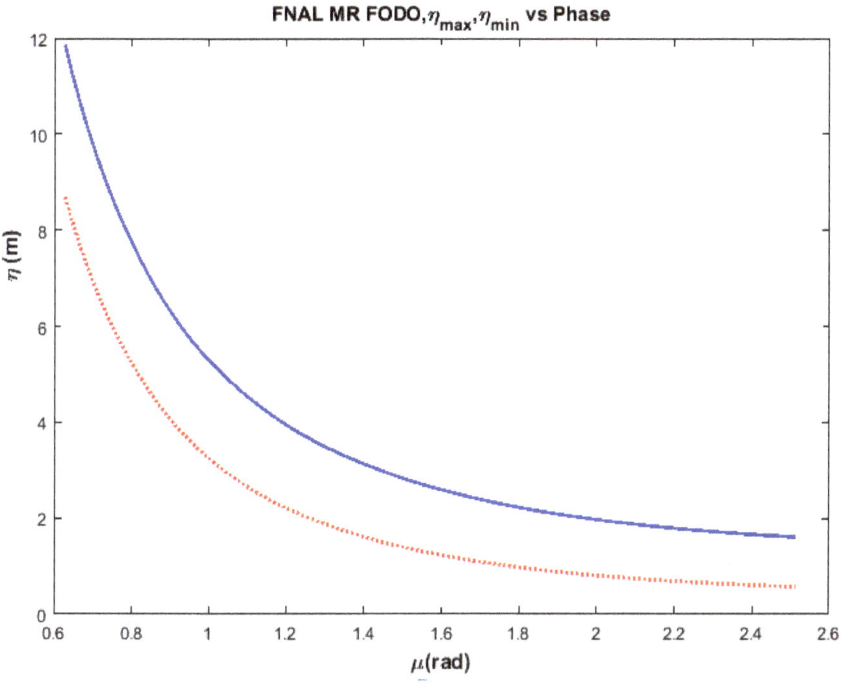

Figure 9.5: Plot of the maximum (solid) and the minimum (dotted) dispersion in a FODO with the bend angle and distance of a FNAL MR FODO as a function of the phase of the FODO.

and β functions for the RHIC FODO are shown in Figure 9.7. The emittance is unchanged by definition but the physical size of the beam is increased with dispersion taken into account to a value $\sigma_x^2 = \varepsilon_x \beta_x + [D(s)\delta]^2$. This places a limit on the momentum spread which falls within the acceptance of the physical vacuum apertures. From the point of view strictly of minimizing dispersion, larger phase advances are favored.

There are many situations where one wishes to make the dispersion η in some region zero. For a beam situation, there might be a sharp focus requested, such as in the pion/kaon separation for the JPARC beamline or a parallel region for Cerenkov identification. One solution, as used in the JPARC beam example, is to employ a sextupole. That element has a gradient which varies with the dispersive beam x offset and therefore compensates for it. The sextupole

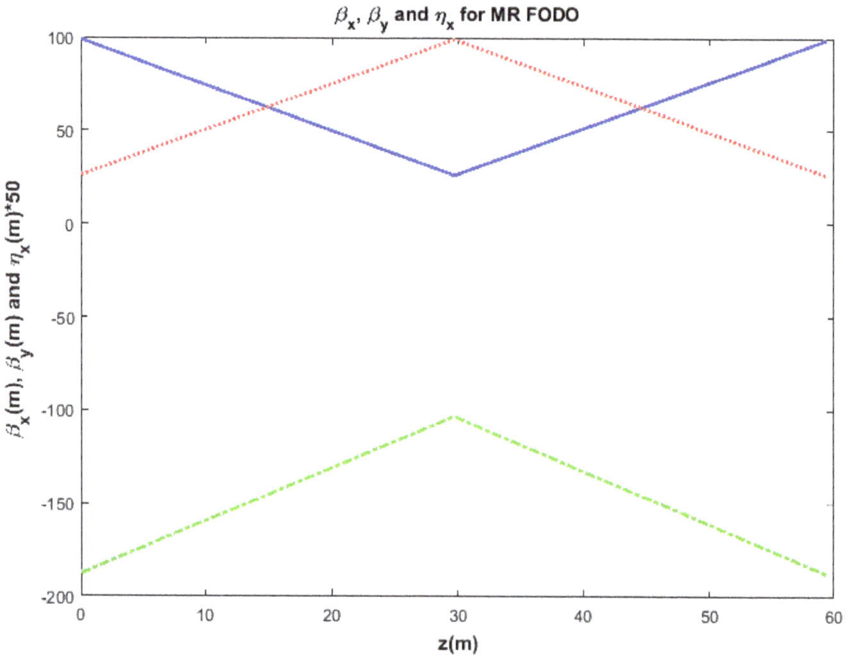

Figure 9.6: FNAL MR FODO β parameters are shown; blue solid for x and red dotted for y. The horizontal dispersion η times -50 is shown with the green dot–dash symbol, which is also a maximum in the F quadrupole.

field is quadratic in the transverse position, $B_s = (d^2B/d^2r)x^2/2$, as mentioned previously in Chapter 7.

Beams without sextupoles can also be designed to have no dispersion, or to be achromatic. That condition would give the smallest focus in the x direction. Alternatively, making the dispersion large gives the possibility of measuring the beam momentum particle by particle, by placing position-sensitive detectors at a point of large dispersion.

A simple beam example with zero dispersion is a layout with two dipole magnets of length d with bend angles ϕ_1 and ϕ_2, separated by a distance $d_1 + d_2$. The distance d_1 is that from the exit of the first dipole to a thin lens quadrupole of focal length f, and d_2 is the distance from the lens to the entrance of the second dipole. This geometry is explored in the script "Beam_Zero_Dispersion".

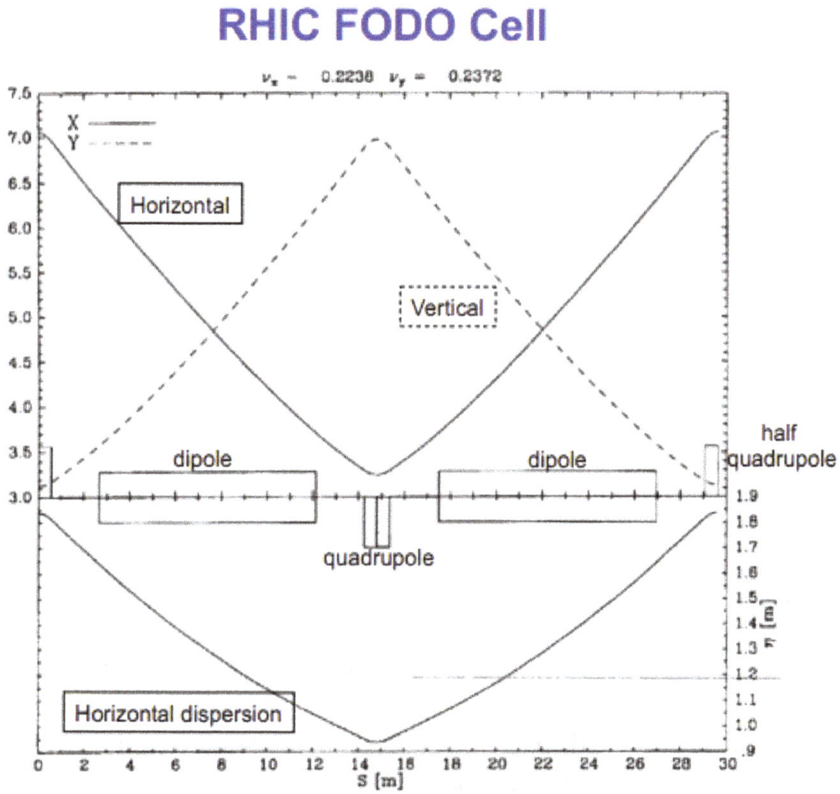

Figure 9.7: RHIC FODO with the square root of β_x and β_y plotted, and the η_x dispersion for comparison with the MR FODO.

A printout of the script appears in Figure 9.8. The full transport matrix is computed and two elements, $D = M(1,3)$ and $D' = M(2,3)$, are shown symbolically. The symbolic solutions for d_1 and d_2 after requiring $M(1,3)$ and $M(2,3)$ to simultaneously be zero are also printed. The utility "solve" is used in this script to find the two distances.

The solution for the two distances as a function of the quadrupole focal length, the drift distance, and the dipole bend angles is

$$\begin{aligned} d_1 &= f - d/2 + f(\phi_2/\phi_1) \\ d_2 &= f - d/2 + f(\phi_1/\phi_2). \end{aligned} \tag{9.8}$$

```
>> Beam_Zero_Dispersion
  a bend, a drift, a quadrupole, a drift, a bend - Matrices and ray traces with dispersion


MT(1,3), MT(2,3) Transfer Matrix
                                                                   / d + d2      \
                                                           d phi1 | ------ - 1 |
d phi2           /                 / d + d2       \ \              \    f        /
------ + phi1 | d + d2 - d1 | ------ - 1 | | - ---------------------
   2             \                 \    f         / /                  2

            / d1      \   d phi1
phi2 - phi1 | -- - 1 | - ------
            \  f      /    2 f

Drift d1 and d2 in Terms of f, phi1 and phi2 for D=Dp=0
2 f phi1 - d phi1 + 2 f phi2
----------------------------
          2 phi1

2 f phi1 - d phi2 + 2 f phi2
----------------------------
          2 phi2

 For phi1 = phi2 = 0.05, f = 30, d = 10 d1 and d2 are 55, 55
```

Figure 9.8: Printout of the script "Beam_Zero_Dispersion".

For equal dipole bend angles the two drift distances are equal: $d_1 = d_2 = 2f - d/2$. A plot of the x ray and both D and D' for that set of parameters is displayed in Figure 9.9. The quadrupole acts as a "field lens" to capture the dispersion. It is then normally going to be the limiting aperture of the beam.

After the defined choice of parameters is executed and plotted, the user is invited to keep the initial beam parameters and the initial bend angles, but to choose a new value of both f and d in order to see what the plot of x, D, and D' values then looks like.

There are other ways to remove beam dispersion. One such is the "dogleg" with two dipoles, but bending in opposite directions with intermediate quadrupoles and drift elements used to remove both D and D'. Removing dispersion in an accelerator FODO is also useful in order to minimize the beam size with a finite dp/p to ease injection of a beam or extraction. There are several methods but only the simplest and most common one is explored here, called the missing magnet solution.

Consider the simplest accelerator case of two FODOs, which have different dipole strengths, ϕ_1 and ϕ_2. The two independent magnet

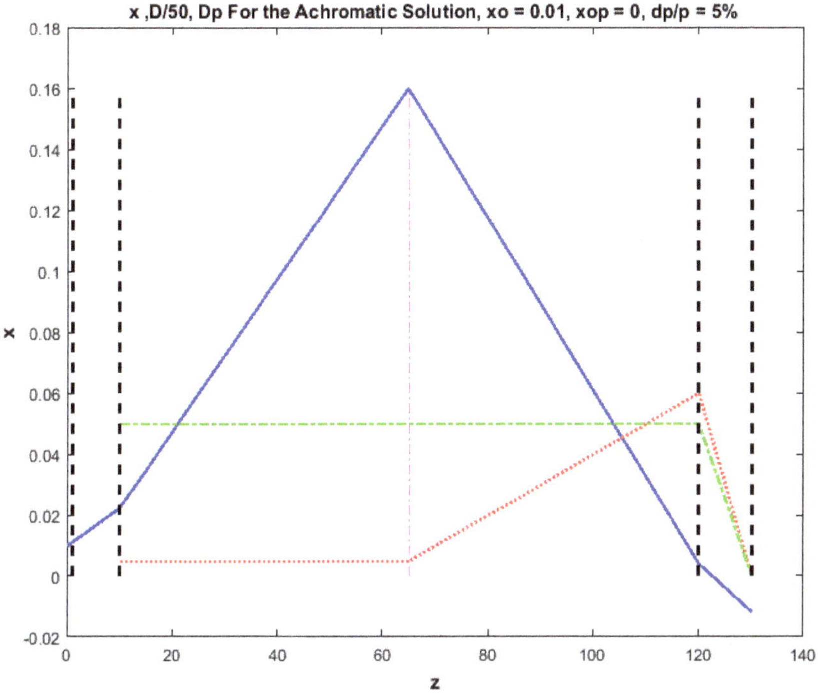

Figure 9.9: Ray trace for the dispersionless beam example. The x value (solid blue) peaks in the focusing quadrupole, as expected. The dispersion, D (red dotted), induced by the first dipole is magnified by the quadrupole and then zeroed out by the second dipole, as is D' (green dot–dash). The dipole boundaries appear as black dashed vertical lines, while the thin quadrupole location is shown as a magenta dash–dot vertical line.

strengths (or better lengths) allow the basic FODO to have the dispersion suppressed. Since only dipoles create dispersion in this approximation, having corrected for dispersion, any system of purely quadrupole elements following the double FODO will also remain dispersionless. The layout is shown in Figure 9.10.

The situation is explored using the script "Dispersion_Zero_2". The dipoles are assumed to fill all the space between the thin, zero length quadrupoles. The transfer matrix for the two FODOs is first multiplied out symbolically and then the condition that the dispersion η and η' be zero, Eq. (9.8), at the exit of the second FODO is imposed and the requirements on ϕ_1 and ϕ_2 are evaluated.

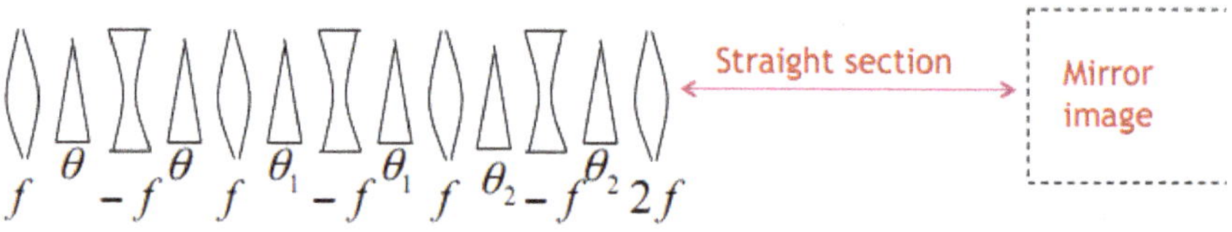

Figure 9.10: Layout of the two FODOs with different dipole magnet sets needed to suppress dispersion in a straight section which follows.

```
>> Dispersion_Zero_2
Full MT Transfer Matrix for the 2 FODO, D = M(1,3), Dp = M(2,3)
Solve for x=0, xp=0 at end of second FODO for xo = eta, xpo = 0 dp = 1
Variables solved for are phi1 and phi2
        2    2
eta (d   - f )
-------------
d f (d + 4 f)

    eta f
-----------
d (d + 4 f)

Solution for (phi1+phi2) ~ eta
    d eta
-----------
f (d + 4 f)

cos2u for the Double FODO
 4       2  2      4
d  - 4 d  f  + 3 f
--------------------
              4
           2 f

For 2 MR FODO eta = 4.9005, phi2 = 0.033, phi1 = 0
```

Figure 9.11: The printout for "Dispersion_Zero_2", showing D and η. The betatron factors for the two FODOs are unchanged.

The printout of the script appears in Figure 9.11. The matrix element for $D = M(1,3)$ and $D' = M(2,3)$ is printed and the solution for no dispersion, which was found using the utility "solve", is also printed. The constraint equations for traversing two unit cells with α and η'

of zero at the cell boundaries are

$$\begin{pmatrix} 0 \\ 0 \\ 1 \end{pmatrix} = \begin{pmatrix} \cos(2\mu) & \beta\sin(2\mu) & D \\ -\sin(2\mu)/\beta & \cos(2\mu) & D' \\ 0 & 0 & 1 \end{pmatrix} \begin{pmatrix} \eta \\ 0 \\ 1 \end{pmatrix}. \quad (9.9)$$

Using the Q_F centerline for the two FODOs means that $\alpha = 0$, $\gamma = 1/\beta$, and $\eta' = 0$. The dispersion suppression constraint is then $M(1,1)\eta + D = 0, M(2,1)\eta + D' = 0$, which can be formulated equivalently in terms not of the elements of M directly, but of the α, β, γ, and μ parameters. The dp/p parameter is taken to be 1, as it was before. It can be inserted later: $D \to D\delta, D' \to D'\delta$.

The previous FODO results for the reference orbit betatron oscillations can be used, such as $\sin(\mu/2) = d/2f, \cos(2\mu) = 1 - 2\sin^2(\mu)$. Since the transfer matrix is for two unit cells, the phase is 2μ by invoking the periodicity of the unit cell. Both formulations will be noted, although the M approach is better matched to MATLAB utilities.

Note that if $d = f$ then ϕ_1 is zero, and no magnets are needed in the first FODO. Hence the term "missing magnet" solution. It is also of interest that, in general, η depends only on the sum of the dipole bend angles. In fact, the numerical evaluation shown in Figure 9.12 has had the MR FODO parameters tweaked very slightly with respect to those shown in Table 9.1, to make the magnets missing by setting f to d. The value for ϕ_2 is the normal MR bend angle for a half cell. The dispersion is seen to monotonically decrease to zero at the end of the second FODO.

After the previous results for phase advance in the FODO are substituted instead of the f and d formulation, the values of the two bend angles that satisfy the constraints are

$$\begin{aligned} &\phi_1 = [1 - (f/d)^2](\phi_1 + \phi_2), \phi_2 = (f/d)^2(\phi_1 + \phi_2) \\ &\phi_1 = [1 - 1/(4\sin^2(\mu/2)](\phi_1 + \phi_2) \\ &\phi_2 = [1/(4\sin^2(\mu/2)](\phi_1 + \phi_2). \end{aligned} \quad (9.10)$$

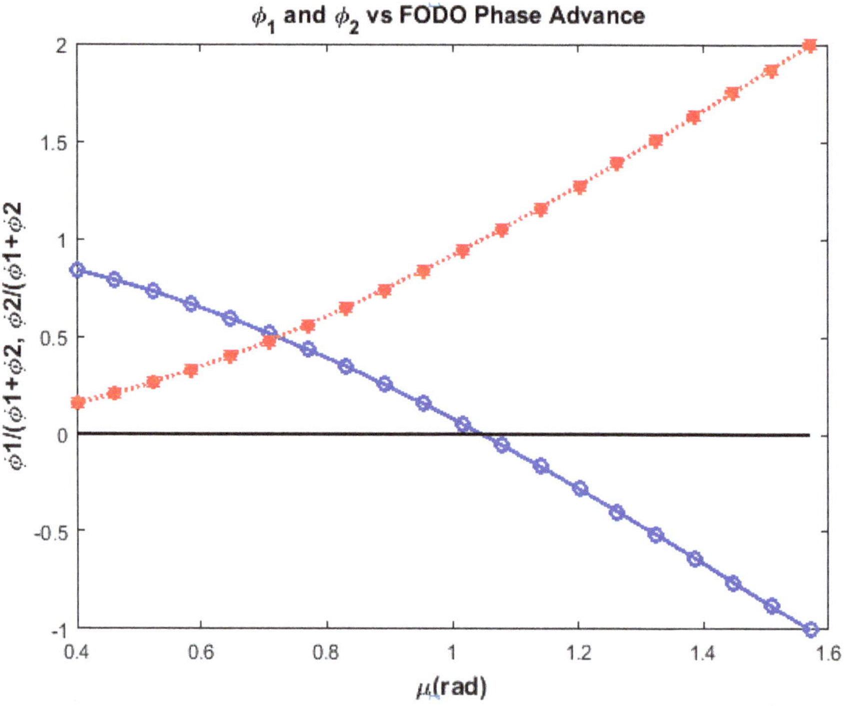

Figure 9.12: Dipole bend angles for the two FODOs as a function of the FODO phase advance for a dispersionless exit. The blue o solid line is for ϕ_1, while the red dotted * line is for ϕ_2. The black horizontal line indicates a missing magnet.

The two bend angles are displayed as a function of the FODO phase advance using the script "Missing_Magnet" and appear in Figure 9.11. Recalling that a small phase advance is disfavored because of the large values of β, the solutions with ϕ_1 small or even missing are favored.

The solution with a missing magnet occurs at a 60° phase advance, $f = d$, for the FODO, and the dispersion to be canceled also follows from the printout of Figure 9.11

$$\eta = \phi_2 d(4 + d/f). \tag{9.11}$$

A plot of the dispersion in the case of a missing magnet solution using the MR as the FODO appears in Figure 9.13.

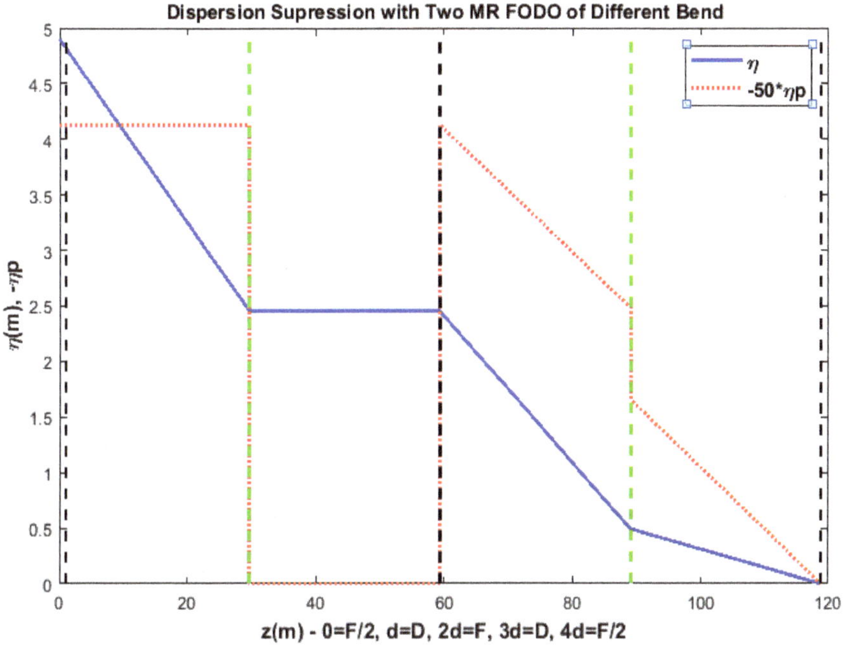

Figure 9.13: Plot of η and η' for two MR FODOs. The second cell has MR dipoles, while the first does not. The MR f has been set to d to ensure that the magnets are, indeed, missing. The black vertical dashed lines indicate the F quadrupoles, while the green dashed lines show the location of the D quadrupoles.

A plot of the beta functions and the dispersion over a wider range of s appears in Figure 9.14 for another unit cell. The dispersion smoothly goes to zero, and with the same pattern, as was seen already in Figure 9.13.

A similar plot for the Fermilab Main Injector optics at injection appears in Figure 9.15. The quadrupoles are represented as vertical figures, while the dipoles lie on the horizontal. The FODO to the left has no dipoles and the dispersion has been removed in those FODOs, leaving room for the needed injection hardware. The dipoles in the last two FODOs are not missing but are shorter than the dipoles in the basic Main Injector unit cells.

The chromatic effects of the quadrupole focal lengths can also be a cause of dispersion, although it is normally small with respect to the dipole magnet dispersion. These effects can also be corrected

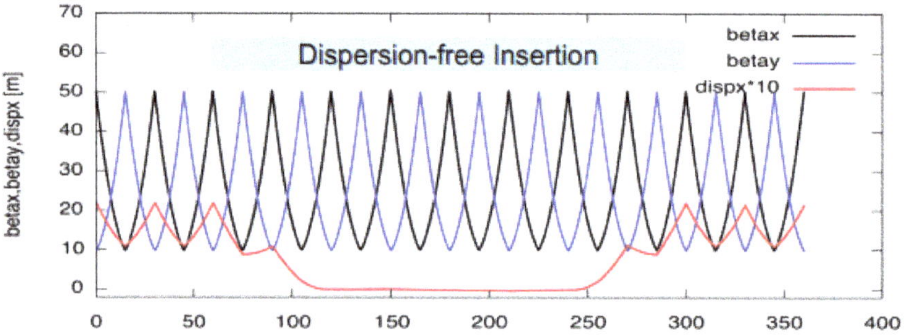

Figure 9.14: Plot of the dispersion using the missing magnet technique. The dispersion-free region contains no dipoles.

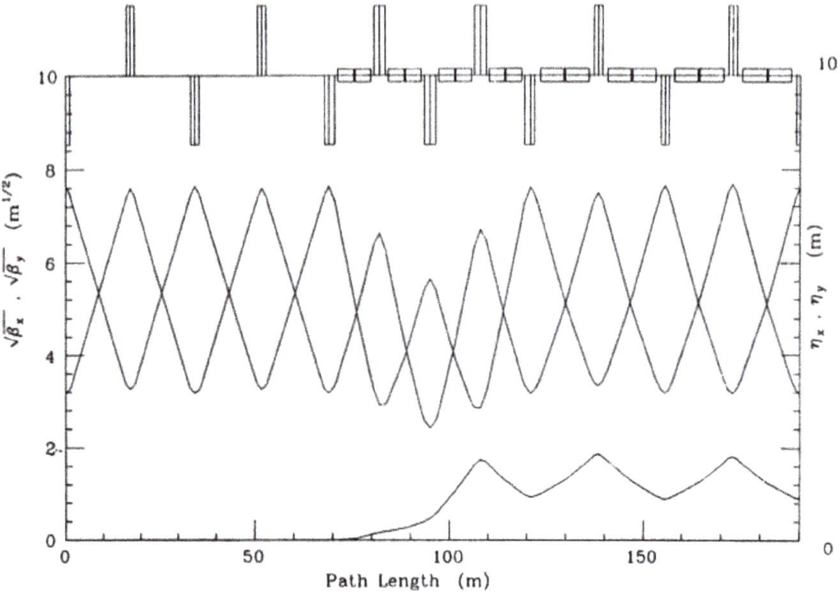

Figure 9.15: Optics at injection for the Fermilab Main Injector. The missing magnet dispersion suppression is quite evident, particularly the reduced dipole length in the FODO before the magnetless FODO. For the Main Injector, the maximum β is 57 m and the maximum dispersion is 2 m.

for by using sextupoles. Assuming one sextupole per FODO with a FODO dispersion η due only to the quadrupole chromaticity, the correction is solved for in the script "FODO_Sext". The sextupole is represented by a thin element at the Q_F center which has a

```
Add sextupole at QF, 1/f kick ~k_s*eta*dp
Shift in FODO phase due to Sextupole
  d dp eta ks (d + 2 f)
- ---------------------
          2 f

Shift in FODO phase due to Quad Chromaticity
    2
  d  dp
- -----
     3
    f

Sextupole ks to Compensate Quad Chromaticity
         2 d
- ----------------
        2
  eta f  (d + 2 f)
```

Figure 9.16: Output of the script "FODO_Sext". The sextupole "kick" is proportional to $\eta\delta$ and contributes a phase shift in the FODO which cancels out the phase shift due to quadrupole chromaticity.

focal length inversely proportional to the dispersion, Eq. (9.2): $1/f_s = -k_s\eta\delta$.

The change in phase of the FODO due to this new element is computed, as is the change due to the chromaticity of the quadrupoles by using the utility "diff" to evaluate the phase change as the focal length changes with momentum. A printout of the solution to make the phase shift through the FODO invariant with the chromaticity, $\delta = dp/p$, appears in Figure 9.16. It uses the "solve" utility to find that $k_s = -2/[\eta f^2(1 + 2f/d)]$. Note that the solution for f_s is independent of η, at least in the lowest order.

Having learned how to cancel the dispersion, the insertion of dipole-free regions or "straight sections" can now be examined. These regions are vital for accelerator options which include beam insertion and extraction, acceleration, and experimental apparatus.

Chapter 10

Insertions

Despite my resistance to hyperbole, the LHC belongs to a world that can only be described with superlatives.

— **Lisa Randall**

The assumption so far has been a regular lattice of unit cells. However, there is a need to accommodate acceleration, injection, extraction, and collisions of beams, as well as other operations. For this reason "insertions" are designed to make these activities possible. However, in order to avoid disruption of the main lattice optics, a "matching" of the betatron oscillations of the insertion and the unit cell is needed. If there is a match, there is minimal disruption and no loss of dynamic aperture in the lattice itself. For example, if the phase advance of the insertion is π the insertion transfer matrix is $-I$ and makes no disruption.

The idea is to match the lattice functions and supply a field-free region where operations may be made. In fact the missing magnet mode is a first step since much of the FODO real estate is taken up by the dipole magnets. For the Collins insertion, the location of the insertion is midway between the F and D quadrupoles of a unit cell FODO. After the unit cell Q_F with f there is a distance $d/2 = L/4$ in Figure 10.1, and then the insertion, characterized by quadrupoles with focal length F, and drift spaces s_1 and s_2. In this text the notation d for the distance between thin quadrupoles in the basic FODO is retained for consistency, so the distances in the script are d, d_1, and d_2. The specific choice of insertion point is chosen so

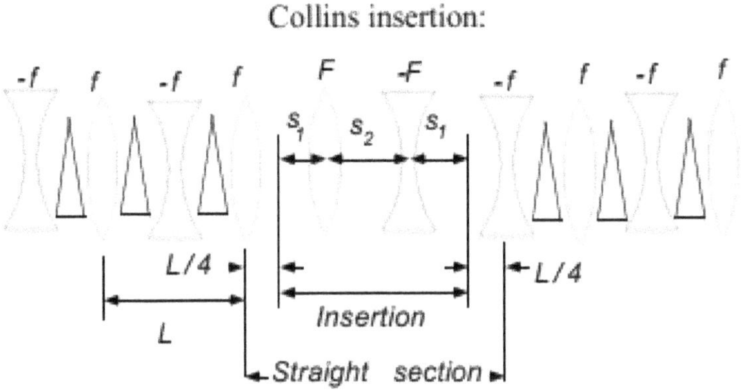

Figure 10.1: Geometry and definitions for the Collins straight section. A free space of length s_2 is provided for the insertion of specialty elements. The notation is that L shown here is $2d$ in the text.

that $\alpha_x = -\alpha_y$ and the matching works in both planes for thin lens FODOs.

The Collins straight section is analyzed in the script "Collins Straight". Initially the transfer matrix for the insertion alone is computed. The matching to the unit FODO lattice is made at the entrance of the insertion. The notation d and f for the unit cell FODO parameters is retained in the script and the insertion drifts are labelled d_1 and d_2 to keep the notation consistent with the body of the text. The transfer matrix is assumed to match to the lattice at the point of insertion defined by the matching lattice parameters, α_m, β_m, and γ_m. The insertion transfer matrix is shown in Figure 10.2.

Examining the overall insertion transfer matrix and using the fact that the transfer matrix defines the Collins lattice parameters, $M(1,1) + M(2,2) = 2\cos\mu$, $M(1,1) - M(2,2) = 2\alpha$, $M(1,2) = \beta\sin\mu$, and $M(2,1) = -\gamma\sin\mu$, the conditions to match the basic FODO lattice parameters, α_m, β_m, and γ_m are $\cos\mu = 1 - d_1 d_2/F^2$, $\alpha_m \sin\mu = d_2/F$, $\beta_m \sin\mu = 2d_1 + d_2 - d_2 d_1^2/F^2$, $\gamma_m \sin\mu = d_2/F^2$.

The largest value of the free space d_2 clearly then occurs when the sine of the phase is 1, which is the reason for the Collins insertion being called a "$\pi/2$ insertion". In this specific case, the parameters of the Collins insertion can be simply related to the

```
>> Collins_Straight
/  2                        2              2          2 \
| F  + d2 F - d1 d2   2 F  d1 + d2 F  - d2 d1         |
| -----------------, ------------------------          |
|          2                      2                     |
|         F                       F                     |
|                                                       |
|                            2                          |
|          d2             - F  + d2 F + d1 d2           |
|       - --,          - -------------------            |
|           2                        2                  |
\         F                         F                  /
```

Figure 10.2: The transfer matrix for the insertion, defined by F, d_1 and d_2.

```
>> Collins_Straight
MT parameters, cosu and alpha*sinu
      2
  - F  + d1 d2
- ------------
         2
        F

d2
--
 F

 Maximum d2 for Straight Section with sinu = 1, pi/2
 F^2 = d1*d2, alpha = d2/F, gamma = -MT(2,1)
d2
--
 2
F

 beta = d1+d2, solving: d1 = 1/gamma, d2 = alpha^2/gamma, F = alpha/gamma
Match at QF end + d/2,  d1 = 20.8552, d2 = 31.4157, F = 25.5965
```

Figure 10.3: Printout of the symbolic transfer matrix and a specific insertion for the MR FODO when the Collins insertion occurs a drift distance $d/2$ away from the F quadrupole, as seen in Figure 10.1. That location is chosen so that the matching works in both x and y.

unit cell parameters at the point of insertion, where the subscript m indicates the matching betatron parameters: $d_1 = 1/\gamma_m$, $d_2 = \alpha_m^2/\gamma_m$, $F = \alpha_m/\gamma_m$.

The symbolic solutions for the Collins insertion are shown in Figure 10.3, as well as the numerical results for creating a Collins insertion for an MR FODO unit cell. A subsidiary script, "ell_prop", is used to start from the Q_F FODO centerline, where $\alpha = 0$, β is the maximum value and γ is $1/\beta$, and propagate the lattice parameters to the matching point and then further through the Collins section, and finally into the remainder of the MR FODO into which the Collins straight section was inserted.

A plot of the beta functions in both x and y for a Collins insertion is shown in Figure 10.4, computed numerically in the script "Collins_Straight". Note that the x and y functions in the insertion are the mirror images on one another. It is that symmetry that makes

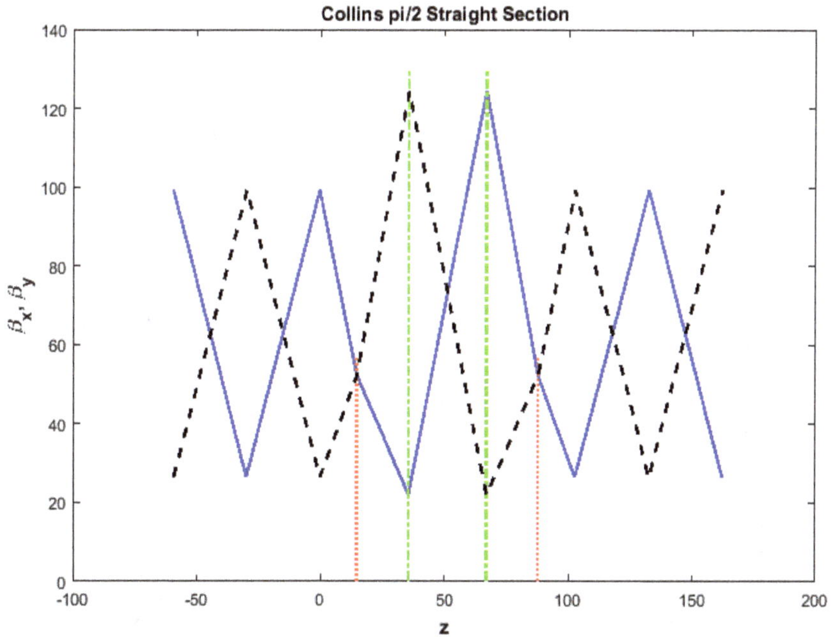

Figure 10.4: Results for a Collins straight section inserted into an MR FODO. The dotted vertical lines indicate the matching points. The dot–dashed points indicate the Collins quadrupoles. The solid line is β_x, while the dashed line is β_y.

the requirement of a $d/2$ insertion point for the Collins nonperturbing in both views. A free space of about 31 m is obtained for experimental use in this example.

It is observed that a free space d_2 slightly longer than d is possible. However, there is a cost. The focal length F is near in value to that of the unit cells and the beta function is larger. This means that the limiting aperture is likely to be in the quadrupoles of the insertion. Keeping the same beam acceptance then requires insertion quadrupoles of larger spatial aperture than the unit cell. This problem is much worse in the case of a "low beta" insertion, which will be explored later.

Optics for the Fermilab Main Injector straight sections are shown in Figure 10.5. The dispersion is corrected to be near zero by using different bend strengths prior to the insertion. The quadrupoles appear as vertical lines, while the dipoles are small rectangles as in Figure 9.15. The insertion increases the maximum value of the betatron envelope.

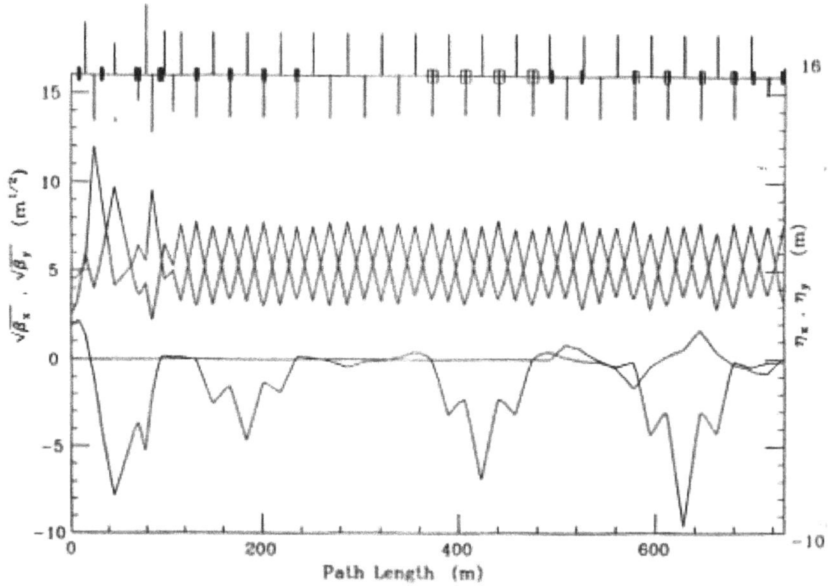

Figure 10.5: Optics at Main Injector straight sections; the square roots of β_x, β_y, and η are plotted in this case. The beta functions at the insertions increase about four times over those in the FODO lattice.

Chapter 11

Low Beta

I always tend to assume there's an infinite amount of money out there.

There might as well be — but most of it gets spent on pornography, sugar water, and bombs. There is only so much that can be scraped together for particle accelerators.

— **Neal Stephenson**

A historical plot of the achieved energy of the fundamental colliding constituents of beams as a function of time appears in Figure 11.1. The data follows a roughly exponential rise with time, although that steep dependence occurs substantially because of the conversion of fixed target accelerator experiments to colliding beams in storage rings. Data is shown both for electron–positron colliders (green circles) and for proton–antiproton and proton–proton colliders (red squares). This energy increase allows an increase in mass for the production of new fundamental particles and this is the essence of the "energy frontier". Some of the new particles found at these facilities are indicated. The Higgs boson discovery at the LHC was announced in 2012.

The growth in the center of mass (CM) energy, labeled m here, occurs for two reasons. First, the dipole magnets have a maximum field which has gone from about 2 T (limit due to iron saturation) to 4 T at the Tevatron to 8 T at the LHC. The fact that the dipoles are now superconducting also allows accelerator facilities to drastically reduce their power bill.

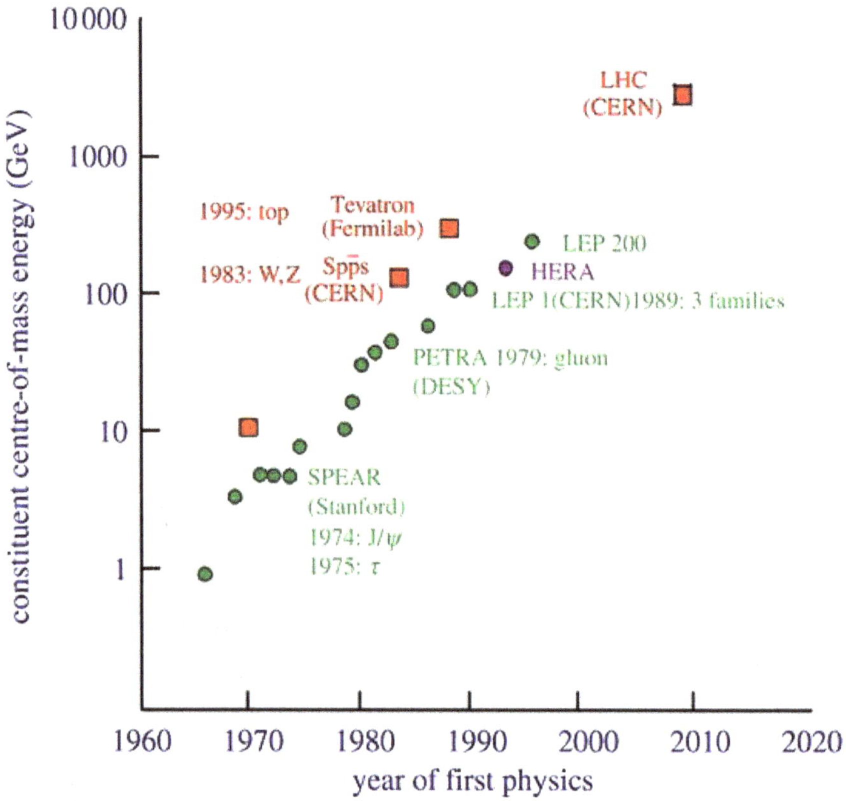

Figure 11.1: Plot of the available energy to produce new particles as a function of time. This is historically called a "Livingston plot". The unlabeled square at year 1970 refers to the CERN Intersecting Storage Rings (ISR), the first hadron collider.

The mass of a system of particles has the same expression as the mass of a single particle, with ε now the sum of the particle energies and p the magnitude of the vector sum of the particles. In the specific case of a two-particle system, there are two limiting cases, shown in Eq. (11.1). In the first case of a beam hitting a stationary target, the available mass goes as the square root of the beam energy, while for the second case of a head-on collision it goes as the colliding energy. For example, the LHC is equivalent to a 98,000 TeV fixed target beam, which is totally not affordable and which shows the great advantage of colliding beams.

$$\begin{aligned} m^2 &= \varepsilon^2 - \bar{p}^2 \\ &= m_1^2 + m_2^2 + 2\varepsilon_1 m_2 \sim 2\varepsilon_1 m_2 \\ &= m_1^2 + m_2^2 + 2(\varepsilon_1\varepsilon_2 - \bar{p}_1 \cdot \bar{p}_2) \sim 4\varepsilon_1\varepsilon_2. \end{aligned} \tag{11.1}$$

However, as the mass of produced systems, m, increases, the probability of producing the system falls rapidly with m. In fact, the cross section for a given mass falls generically as the inverse square of the mass which has the correct dimensions ($\hbar$). In colliding beams the rate of the production of a mass m depends not only on the physics — the cross section — but also on the luminosity. If a mass a factor of 2 above what is currently known is to be explored, the luminosity will need to be increased by a factor of 4 to achieve the same rate of production.

$$R = \sigma\ell. \tag{11.2}$$

In colliding beam applications, the reaction rate, R, or the luminosity, ℓ, is maximized by reducing the beams that are colliding to the smallest dimensions and the highest density.

$$\ell = N_1 N_2 f n_b / 4\pi\sigma_x\sigma_y. \tag{11.3}$$

The luminosity depends on the number of particles in a "bunch" (shaped by the RF acceleration, as will be discussed later) as N_1N_2, the number of bunches, n_b, and by the frequency f with which they complete a turn/revolution of the accelerator. The factor $N_1N_2fn_b$ is easy to visualize. For a particle in bunch 1 there are N_2f crossings per second and there are N_1 such particles in each of n_b bunches. The transverse density for Gaussian beams, with x and y characterized by RMS sizes, σ_x and σ_y, appears in the denominator for the luminosity which specifies the transverse density appropriate to the collisions.

For the LHC, with $N_1 = N_2 = 1.1 \times 10^{11}$, $n_b = 2.8 \times 10^3$ the rotation frequency of the 27 km ring is $1.1 \times 10^4\,\text{sec}^{-1}$. The low beta focus of the beams has an RMS size of $17\,\mu$m. Those parameters

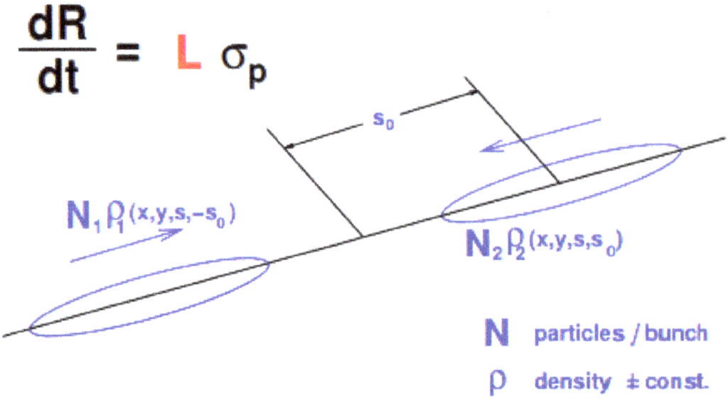

Figure 11.2: Schematic diagram for an LHC collision region with head-on proton–proton collisions.

lead to a luminosity of $1.07 \times 10^{34}/\text{cm}^2{*}\text{sec}$. For a total cross section of 100 mb (10^{-25} cm^2), the full reaction rate is about 1 GHz. A schematic of an LHC collision region is shown in Figure 11.2.

There are limitations on increasing the luminosity very much beyond this value. For example, there are limits on how many particles can be crammed into a "bunch" since they repel each other, as will be briefly explored later. There are also limits imposed on the magnets used to minimize the beam sizes at the collision point. The luminosity scales as the inverse square root of the product $\beta_x\beta_y$, as follows from the previous discussion of the beam ellipse parameters, $\sigma = \sqrt{\beta\varepsilon}$. A very low value of β at the collision point implies strong focusing, which in turn means large lens sizes needed to capture the rapidly diverging beam growing from the interaction point (IP).

The history of the luminosity evolution at hadron colliders is displayed in Figure 11.3. From the ISR to the later LHC the luminosity growth has been impressive. In general, the proton–antiproton colliders suffer in comparison with the proton–proton colliders, owing to the necessity to create and store antiprotons. The LHC has now (2017) achieved its design luminosity, which puts it at the top of the graph of Figure 11.3.

The jargon for maximizing the luminosity is to form a "low beta" insertion to optimize by creating the smallest practical transverse beam sizes. A typical strategy is to combine dispersion suppression

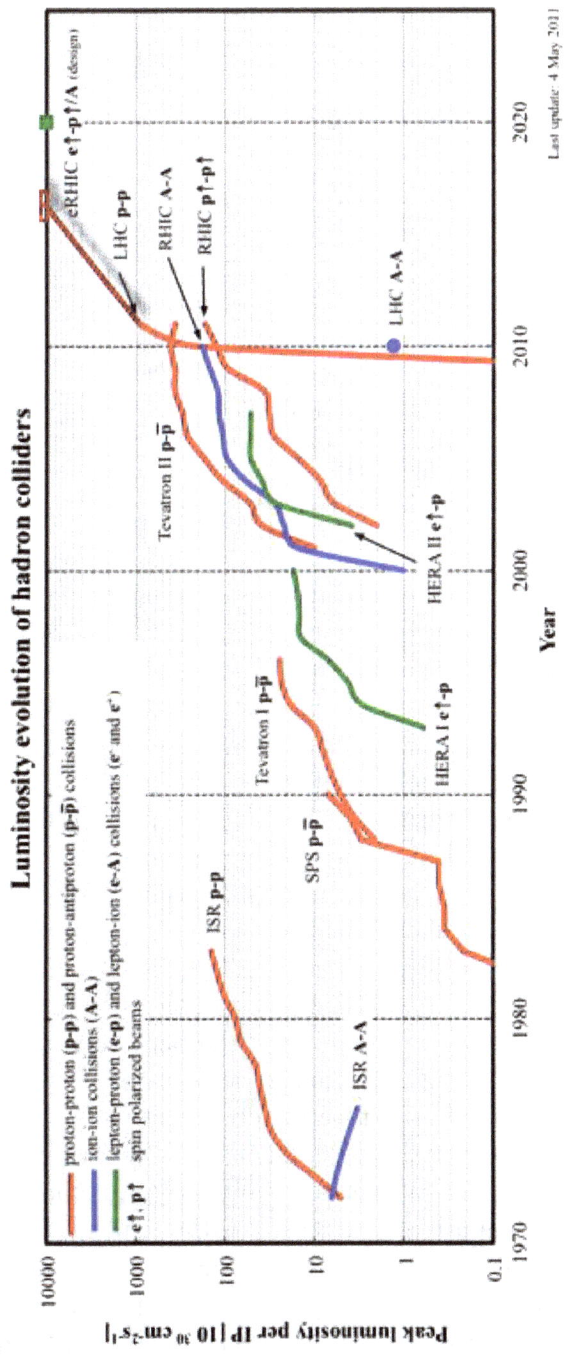

Figure 11.3: Evolution of the luminosity for hadron and electron–proton colliders.

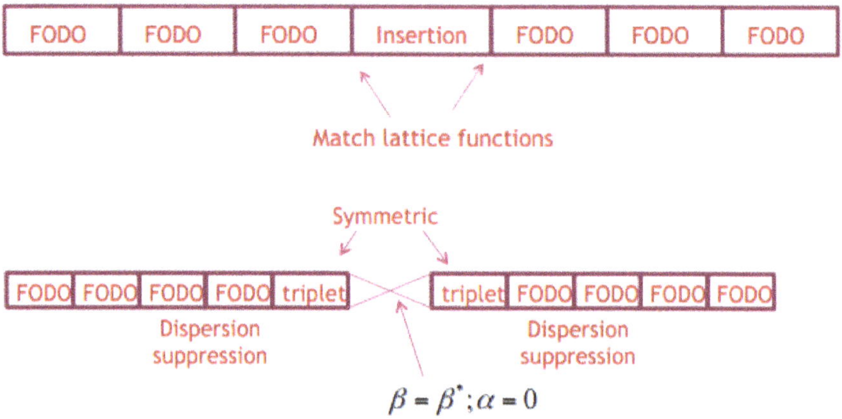

Figure 11.4: Schematic of a low beta insertion, with matched lattice functions, dispersion suppression, and maximum luminosity.

with the insertion of a long straight section. For maximum luminosity the low beta focus should be made in both transverse planes at the same point, because it is the transverse density of the colliding beams that is important. That fact then argues for "round" beams.

A schematic of a typical low beta insertion is shown in Figure 11.4. The main idea is to always match the lattice parameters to avoid large beam excursions within the lattice. Secondly, avoid chromatic beam broadening by combining an initial dispersion suppression in the main lattice with a focusing array in the low beta section, which gives a simultaneous focus in the two transverse planes. This can be achieved by using a triplet of quadrupoles. A "waist" is created at the collision point with parameters β^* and $\alpha^* = 0$. The design plan is to minimize the beta functions in order to maximize the luminosity and hence the physics.

As a preliminary, a short script, "Beta_in_Drift", is used to evaluate the evolution of the β function in a field-free region. The symbolic output appears in Figure 11.5. It is obvious that the desire for a small β^* causes a large excursion in the triplets. In fact, it is likely that the triplets will be the limiting apertures. Indeed, this is the case at the LHC, and the high luminosity improvement program at the LHC will necessitate the construction of larger apertures

```
>> Beta_in_drift
beta After a Drift d - initially betas
In the Drift, alpha = 0 and gamma = 1/beta

b =

bs + (d*conj(d))/bs
```

Figure 11.5: Output of the script "Beta_in_Drift", illustrating the quadratic increase in the β function as a function of drift distance.

and quadrupoles with stronger field gradients. The dependence is approximately $\beta(d) \sim d^2/\beta^*$.

From this dependence the phase advance over a drift distance of total length $2d_{\beta*}$ can be computed:

$$\psi_{\beta*} = \int_{-d_{\beta*}}^{d_{\beta*}} ds/\beta(s) = 2\tan^{-1}(2d_{\beta*}/\beta_*). \tag{11.4}$$

Since β^* is made small by construction, the phase advance in the low beta field-free region is almost π. Therefore, the overall insertion contributes the negative of the unit matrix, and insertion matching happens almost naturally, independent of the exact properties of the triplet, because of the long drift spaces in the insertion. The phase advance in the triplet quadrupoles is small because the β function is large there. Each low beta insertion contributes π to the total tune, whereas the Collins insertion contributes only $\pi/2$.

A simplified low beta consisting of two focusing lenses, analogous to a microscope in classical optics, is explored in the script "Simple_Lo_Beta". The low beta is inserted in the lattice at a location where $\alpha = 0$, for example at a quadrupole. A lens with focal length f_1 matches the lattice while, after a drift of d_{lb}, the strong lens, f_2, focuses to a point d_{bs} at the smallest beam size. These two lenses can be thought of as the effective focal lengths of a triplet or a doublet, for example. Only one dimension is treated in this way, while a true design must function properly for both transverse dimensions. The printout of "Simple_Lo_Beta" appears in Figure 11.6.

```
>> Simple_Lo_Beta
Simplified low beta - f1 - dlb - f2 - dbst
/                 / dbs      \                                    \
|      dbs - dlb | --- - 1 |                                      |
|                 \  f2      /    dbs                / dbs      \ |
| 1 - --------------------- - ---, dbs - dlb | --- - 1 | |
|                   f1                f2                \  f2      / |
|                                                                  |
|               f1 - dlb + f2                          dlb        |
|           - -------------,                       1 - ---        |
\                   f1 f2                              f2         /

Focal Lengths for MT = -I
      2
   dlb
---------
dbs + dlb

 dbs dlb
---------
dbs + dlb

MT for Diagonal Solution, M11 is Demagnification
/    dbs         \
| - ---,    0    |
|    dlb         |
|                |
|           dlb  |
|    0,    - --- |
\           dbs  /
```

Figure 11.6: Output of the script "Simple_Low_Beta" with the half insertion transfer matrix, the focal lengths for the two lenses when the total for the full insertion is $M = I$ is required, and the half transfer matrix in that special case.

The focal lengths are determined — using "solve", as usual — by requiring that the system force the full transfer matrix to be $-I$ matrix, $M(1,2) = M(2,1) = 0$, insuring that the full low beta has $M = -I$ and makes small lattice disturbances. The overall advance is π for the full, symmetric low beta insertion.

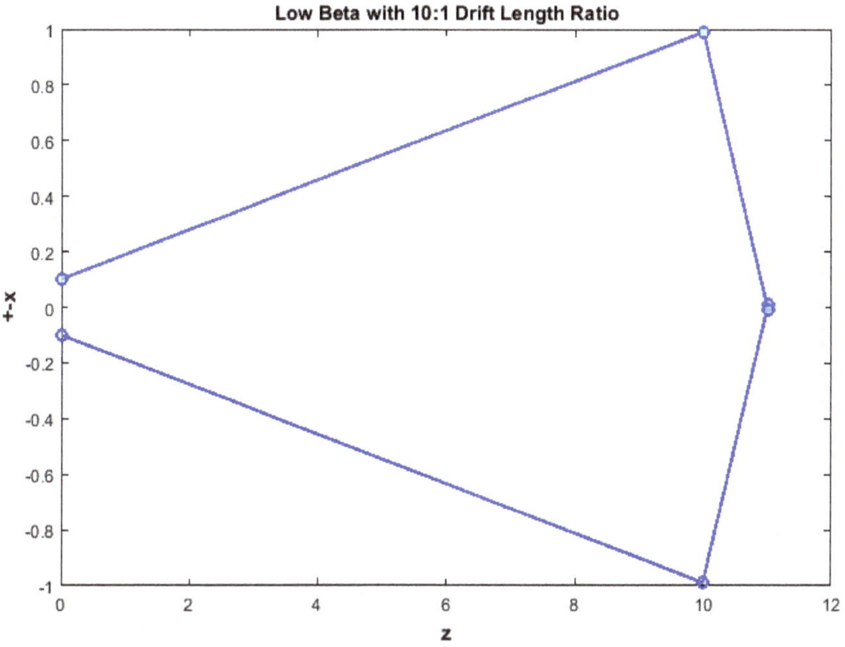

Figure 11.7: Simplified low beta ray for a set of two lenses with a distance between them 10 times the distance of the second to the low beta focal point.

The lattice value, $\beta(s)$, transforms simply since the insertion M is diagonal: $\beta_* = \beta(s)M(1,1)^2 = \beta(s)(d_{lb}/d_{\beta^*})^2$. This result confirms that a high luminosity requires a long straight section, with $d_{lb} \gg d_{b*}$. The strong focusing then means that the beam envelope is large in the f_2 quadrupole, as shown in Figure 11.7, where ray traces are displayed. The approximate values for the focal lengths are $f_1 \sim d_{lb}$ and $f_2 \sim d_{\beta *}$. The magnification is approximately $d_{l\beta}/d_{\beta *}$.

The two-lens system is a reasonable approximation. However, a simple single triplet in the low β insertion is also worth exploring, as is done in the script "Low_Beta_Triplet". The transfer matrix is calculated symbolically and the quadrupole strength is solved for under the constraint that $M(1,1)$ for the transfer matrix vanishes in both transverse planes assuming a parallel beam as input to the insertion. The script is similar to that used to generate Figure 6.8.

```
>> Low_Beta_Triplet
  thin lense symmetric triplet - low beta round beams


Distance to Q1: 5
Distance Q1 to Q2 CL = Distance Q2 to Q3 CL: 5
Distance Q3 CL to Low Beta Focus: 10
 The (x,z) and (y,z) doublet matrices, MT(1,1), NT(1,1)
                                 2        2           2       2                      2
(2 d dlo - 2 d fF - 2 dlo fF + fF ) fD - d  fF + dlo d  + d fF  - dlo d fF 2 + dlo fF
--------------------------------------------------------------------------------------
                                        2
                                      fF  fD

                                 2        2           2       2                      2
(2 d dlo + 2 d fF + 2 dlo fF + fF ) fD - d  fF - dlo d  - d fF  - dlo d fF 2 - dlo fF
--------------------------------------------------------------------------------------
                                        2
                                      fF  fD

 Symbolic Solution for fF , fD
 Numerical Solution for fF -3.73658, fD, -4.09849 M11 = M22  = 0, thin
 Numerical Solution for fF -3.67797, fD, -4.14227 min sqrt(bx*by) = 1.21166 |
 sqrt(betx*bety) initial 52.0955, and at low beta focus 1.21166
```

Figure 11.8: Printout for the low beta triplet in the parallel-to-point configuration. In this script d_o is the distance to Q_1 from the initial beam, d is the distance between all quadrupoles, and d_{lo} is the distance from the triplet exit to the low beta focus.

The utility "solve" is used to require parallel-to-point focus in both transverse planes. The printout of the solution appears in Figure 11.8.

This ray tracing by itself does not yield a figure of merit for the low beta, because it imposes conditions on the transfer matrix and not the beams themselves. To maximize the luminosity, the procedure would be to minimize $\sqrt{\beta_x \varepsilon_x \beta_y \varepsilon_y}$ [see Eq. (11.3)). To find that factor, a computation of the ellipse parameters for an MR FODO is made. The insertion point directly follows the Q_F quadrupole propagating from the centerline to the exit point, using the previously discussed ancillary script "ell prop". The triplet script itself employs the MATLAB utility "fminsearch" to solve for the symmetric triplet focal lengths which minimize $\sqrt{\beta_x \beta_y}$ using a numerical script "low_beta_eval". The initial value in the FODO is reduced by a factor of about 43, as shown in Figure 11.8. The parallel-to-point solution is a good first approximation to the

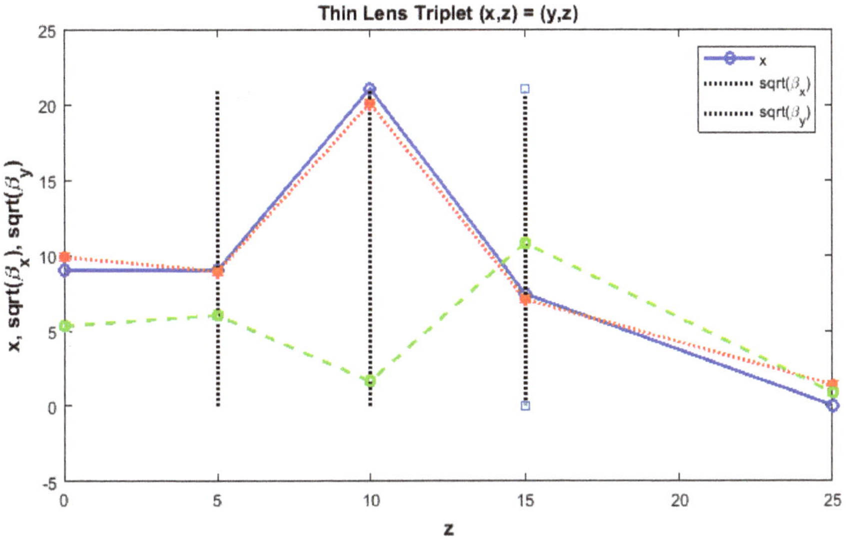

Figure 11.9: Plot of rays in the x direction for a parallel beam exiting the FODO lattice, shown in solid blue. The triplet quadrupole locations are shown as the vertical lines. The beta functions in x (red dotted) and y (green dashed) which match to the FODO at a distance of 5 m beyond the quadrupole are shown for comparison.

optimized solution. Other values can be found using the script, since the user supplies the geometry of the triplet in response to the "menu". There is clearly a tradeoff, as more room for experimentation with the same luminosity necessitates triplet quadrupoles of increased aperture, and therefore expense and technical challenge.

The ray trace in the case of an initially parallel beam which is oval is displayed in Figure 11.9. It is notable that the focus point is approached quite symmetrically. The ray trace in the parallel to point approximation is the starting value for the minimization. The plot displays the square roots of the beta functions in x and y.

A more sophisticated tracking of the beam envelope in a triplet is shown in Figure 11.10. The remarks made previously about the results from the thin lens approximation when comparing the

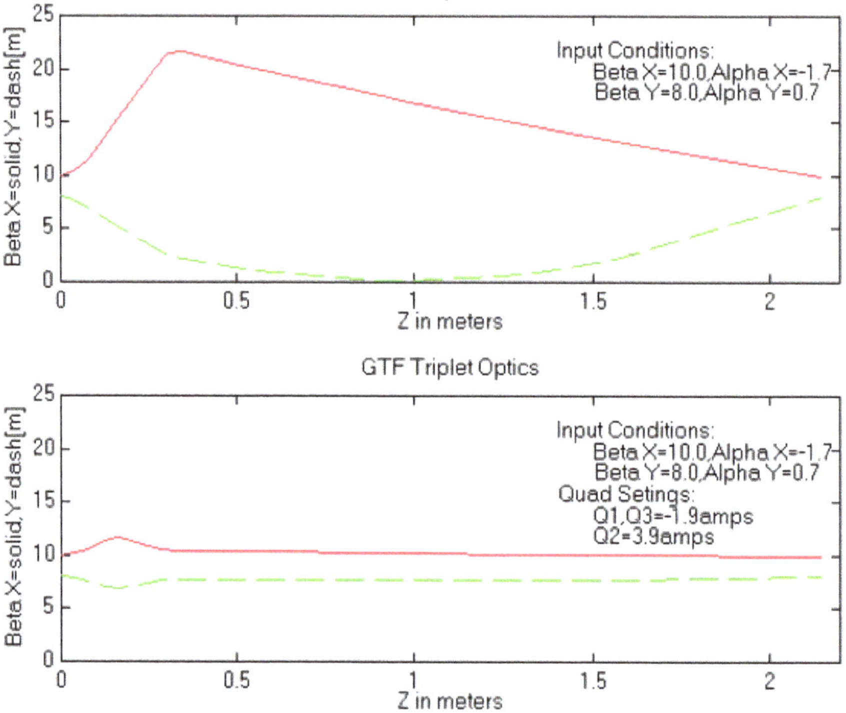

Figure 11.10: Triplet beta functions for specific initial conditions which are matched to yield a round beam of equal size simultaneously in x and y.

quadrupole doublet in both thin and thick lens approximation also hold in this case.

A more realistic low beta insertion for the Tevatron is shown in Figure 11.11. There is some local disturbance of the lattice functions when a high luminosity is achieved. The layout of a low beta region at the LHC is shown in Figure 11.12, where the final focus triplet is evident.

As a final example of a low beta insertion, the RHIC collider at the Brookhaven National Laboratory has a lattice and low beta optics, which are shown in Figure 11.13. Note the absence of dipoles, used for dispersion suppression. Note also the large excursions induced in the low beta insertion. In this plot the beta function is plotted and not the square root, which makes the excursions appear more extreme.

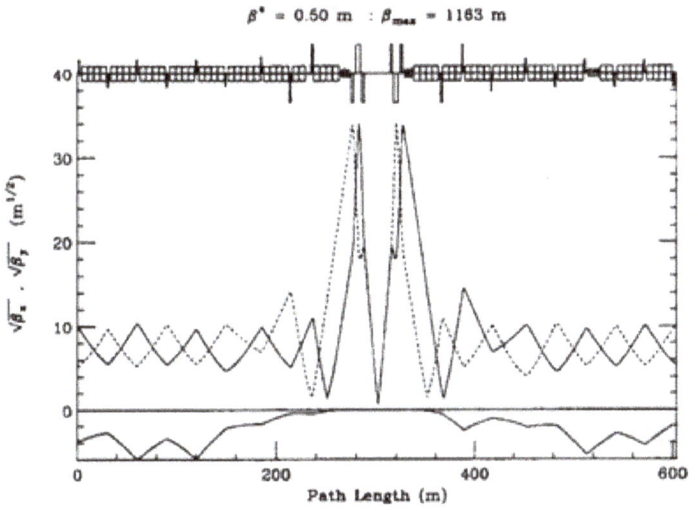

Figure 11.11: Low beta plot for a Tevatron insertion. The value of β^* is reduced to 0.5 m but the maximum envelope increases from about 10 to about 34. The dispersion suppression in the FODO and the small dispersion in the collision region are evident.

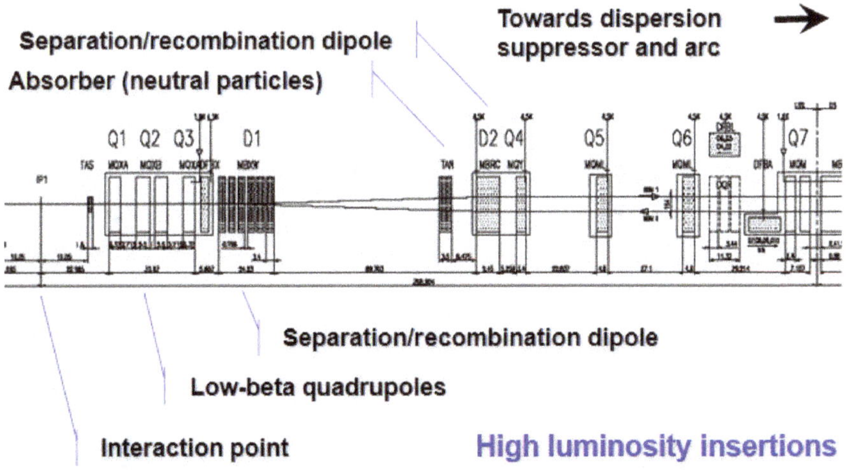

Figure 11.12: LHC low beta region for the ATLAS experiment. The interaction point is shown as IP1, followed by the focus triplet and dipoles to separate the two proton beams. The beams are then in the two beam pipes and matched to the LHC unit cell lattice.

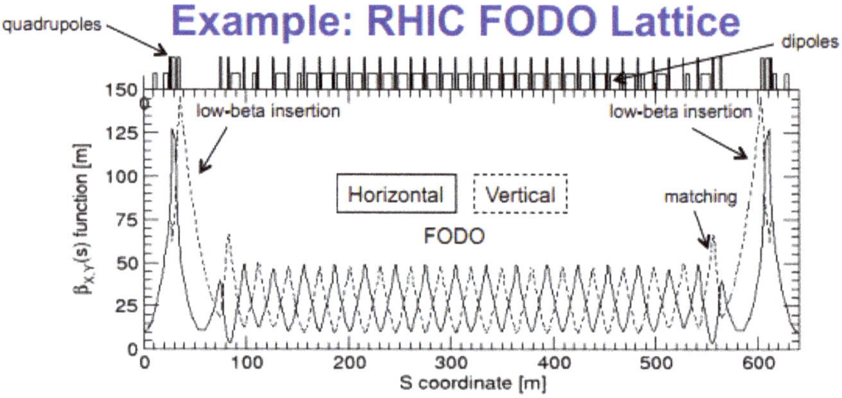

Figure 11.13: Optics for the RHIC lattice and the low beta experimental insertion.

The desire for smaller β^* leads to larger apertures and stronger quadrupoles in the straight sections. For the LHC luminosity increase, the basic superconductor has to be changed from niobium titanium to niobium tin, because the latter will support larger magnetic fields.

Chapter 12

Magnet Imperfections and Beam Interactions

Dr. Egon Spengler: "There's something very important I forgot to tell you."

Dr. Peter Venkmen: "What?"

Dr. Egon Spengler: "Don't cross the beams."

Dr. Peter Venkman: "Why?"

Dr. Egon Spengler: "It would be bad."

— **Ghostbusters**

This piece of engineering is a miracle. It is the pyramids of our time.

— **Frank Close**

The text so far has assumed that the magnets are perfect and the particles in the beam respond only to externally imposed electric and magnetic fields. However, in the real world the magnets are subject to survey errors (translations and rotations) and errors in the fields themselves due to manufacturing tolerances. A special case of "error" was the prior treatment of off-momentum particles and dispersion for which an additional beam parameter, dp/p, and a third dimension were introduced.

In the chapter on "Periodic Structures", the solutions for the equilibrium orbit were established. A simplifying set of assumptions was that there are N unit cells, that the circumference is $C = 2\pi R = N$ times the length of a unit cell FODO, L, and that the phase advance of a unit cell is μ. The total phase advance is then $N\mu$ and the "tune" Q is defined to be $N\mu/2\pi$ or the number of betatron oscillations per turn. The average value of β is estimated to be

$$\begin{aligned} Q &= N\mu/2\pi = N/2\pi \int_o^L ds/\beta(s) \sim NL/2\pi\langle\beta\rangle \\ \langle\beta\rangle &= C/2\pi Q = R/Q, \quad C = NL. \end{aligned} \tag{12.1}$$

For example, in the MR FODO the phase advance per FODO cell is about 71° and the radius is 1000 m. The Q value is about 19.4, so the estimated beta mean is about 52 m. The approximation is somewhat crude but is a plausible first estimate.

First, a followup is made to derive Eq. (8.12) and the more general case. The starting point is Eq. (8.3) for the position as a function of β and the phase ψ. The use of MATLAB in the script "General_Lattice_Point" makes the algebra quite easy. The output of the script appears in Figure 12.1.

First, x at $s = 0$ is differentiated so that, using x_o and x'_o the sin and cos of $\psi_o + \delta$ at that point are determined. Then the cos and sin of $\psi(s) + \delta = \psi_o + \delta + \Delta\psi$ are evaluated. Collecting terms, $x(s)$ and $x'(s)$ are found in terms of x_o and x'_o, β and α at $s = 0$ and s and the sin and cos of the phase change $\Delta\psi$. These final expressions

```
>> General_Lattice_Point
  find transfer matrix in terms of lattice param
  for arbitrary initial and final locations


 Take so = 0 and A = 1 in x(s) = A*sqrt(beta)cos(psi(s)+del)
 Use dpsi/ds = 1/beta and alfa = -(1/2)dbeta/ds
 Differentiate x to find cos(phio+del) and sin(phio+del)
 cos and sin phi + del with phase advance dphi
 Arbitrary Transfer x and xp to lattice b, a with phase advance dphi
sqrt(b) (xo cos(dphi) + ao xo sin(dphi) + bo xop sin(dphi))
-------------------------------------------------------------
                          sqrt(bo)

-(xo sin(dphi) + a xo cos(dphi) - ao xo cos(dphi) - bo xop cos(dphi) + a ao xo sin(dphi) + a bo xop

   sin(dphi))/(sqrt(b) sqrt(bo))

 Special Case - alpha = o in so and s, QF or QD
sqrt(b) (xo cos(dphi) + bo xop sin(dphi))
-----------------------------------------
                 sqrt(bo)

  xo sin(dphi) - bo xop cos(dphi)
- -------------------------------
          sqrt(b) sqrt(bo)
```

Figure 12.1: Output of the script "General_Lattice_Point". The user can examine all the intermediate steps in full detail as desired.

are a bit unwieldy, and are available in the script but are not fully printed out. The full matrix appears in Eq. (12.2). The special case, where α_o and $\alpha(s)$ are zero, the Q_F or Q_D locations, is evaluated and printed out, and Eq. (8.12) is recovered. The result when $\alpha = \alpha_o$, $\beta = \beta_o$ and $\Delta\psi = \omega = 2\pi Q$, which means a full turn for the transfer matrix, is Eq. (12.2), which recovers Eq. (9.6). The symbolic logic tools in MATLAB make the derivation quite simple.

$$\begin{aligned}
&M(s|s_o) \\
&\quad = \begin{bmatrix} (\sqrt{\beta/\beta_o})[\cos(\Delta\psi) + \alpha_o \sin(\Delta\psi)] & \sqrt{\beta\beta_o}\sin(\Delta\psi) \\ [(\alpha_o - \alpha)\cos(\Delta\psi) & (\sqrt{\beta_o/\beta})[\cos(\Delta\psi) \\ \quad -(\alpha\alpha_o + 1)\sin(\Delta\psi)]/\sqrt{\beta\beta_o} & \quad -\alpha\sin(\Delta\psi)] \end{bmatrix} \\
&M(s|s_o) \to M(s+C|s) \\
&\quad = \begin{bmatrix} [\cos(\omega) + \alpha\sin(\omega)] & \beta\sin(\omega) \\ -[(\alpha^2+1)\sin(\omega)]/\beta & [\cos(\omega) - \alpha\sin(\omega)] \end{bmatrix} \qquad (12.2) \\
&\Delta\psi = N\mu = 2\pi Q = \omega, \quad \alpha = \alpha_o, \quad \beta = \beta_o.
\end{aligned}$$

It is assumed that there is a designed reference orbit, (0,0), such that a particle on that orbit remains on it. Other particles will oscillate about the closed orbit but are constrained to be within the lattice phase space ellipse. In the presence of errors there is still a closed orbit but a new one instead of the designed closed orbit. Assume that a dipole field error or a quadrupole misalignment causes an angular "kick" θ and that the kicks are small. In the quadrupole case the kick would be dx/f, where dx is the design beam centroid location at the misaligned quadrupole center. In the magnet case the differential kick due to a magnetic field error would be $\phi_B(dB/B)$.

From an examination of Eq. (12.2), the change in position is $\delta x = (\sqrt{\beta\beta_o}\sin\Delta\psi)\theta$, which is largest in an F quadrupole a phase shift $\pi/2$ away from the misalignment. The new closed orbit, with M the transfer matrix shown in Eq. (9.6), then satisfies

$$I\begin{bmatrix} x \\ x' \end{bmatrix} = M(s+C|s)\begin{bmatrix} x \\ x' \end{bmatrix} + \begin{bmatrix} 0 \\ \theta \end{bmatrix}. \qquad (12.3)$$

```
>> First_Order_Res
      b th sin(om)
 - ---------------
    2 (cos(om) - 1)

 th (cos(om) + a sin(om) - 1)
 ----------------------------
        2 (cos(om) - 1)
```

Figure 12.2: Output of the solutions for the new closed orbit obtained using the "solve" utility.

The short script "First_Order_Res" is employed to solve Eq. (12.3) for the new closed orbit using the "solve" utility and invoking the fact that the lattice parameters are unchanged to the lowest order. The results for x and x', with total phase advance $\omega = N\mu = 2\pi Q$, appears in Figure 12.2.

Using trig identities, (Appendix D), it is clear that there is resonant behavior for integer values of ω times 2π or for Q an integer. For example, the x of the new closed orbit simplifies to

$$x_\theta = -\theta\beta\cos(\omega/2)/2\sin(\omega/2). \tag{12.4}$$

This resonant behavior can be visualized schematically, as seen in Figure 12.3. If the "kick" is applied at the same point on the ellipse for each turn, then the angles will diverge.

Since the location of the errors and their severity are not known, a statistical argument leads to a rough approximation for the total average error due to the N individual errors of

$$\langle x_\theta\rangle = \langle\theta\rangle\langle\beta\rangle\sqrt{N}/2. \tag{12.5}$$

The stochastic nature with uncorrelated errors is reflected in the square root dependence on the number of magnets. This kind of dependence was found in studying the muon beam in prior work in the text. The individual magnets with fractional RMS error θ combine to make an excursion, a limiting aperture, which is larger than than individual errors.

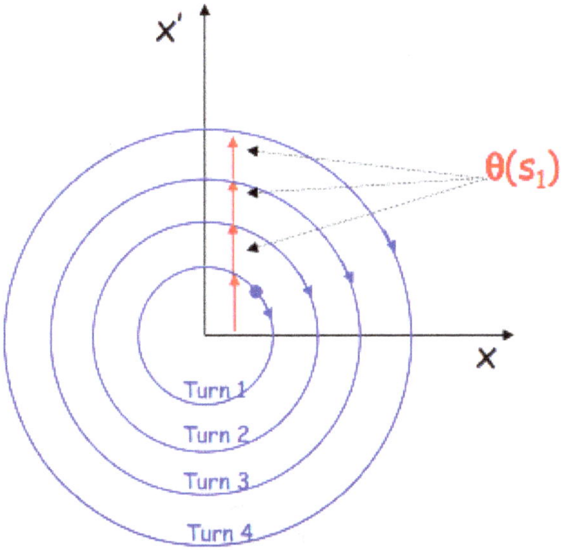

Figure 12.3: Schematic representation for the resonant behavior in the case of a "kick" which occurs at the same point on each turn. The kicks reinforce if they are in phase and the particle diverges rapidly.

For the Main Ring a 1 mm alignment error, with a "kick" dx/f, for 100 quadrupoles implies an excursion of approximately 1 cm. Typically, values of order mm for single magnet alignment errors then translate into an average excursion of order cm. Turning to dipole errors, the shift in x due to a magnetic field error is, from Eq. (12.5), $\beta\phi_B(dB/B)\cos(\omega/2)/\sin(\omega/2)$. Taking MR values for β in Q_F and ϕ_B and a 1% error on the field, and ignoring the trig terms, there is a 3 cm shift for a single dipole with a 33 mrad bend angle.

In order to get some feeling of what the imperfections do if they occur at a single point, a short script named "Magnet_Err" can be used. A FODO is the unit cell which repeats about five unit cells, $Q = 5.1$, as in the MR. Output from the script is shown in Figure 12.4.

Particles are placed on the unperturbed MR ellipse at the F quadrupole center using the random number generator, "rand". Then

```
>> Magnet_Err
  inject dipole error or gradient error and
  propagate thru 8 FODO with dipole error or gradient error


Err Angle, scale 1/sqrt(betamx)~ 0.01: 0.05
Rays in Ellipse
Number of rays to Plot: 20
Err Gradient df, scale fo = 25.6,1/fo + 1/f: 500
Rays in Ellipse
Number of rays to Plot: 20
```

Figure 12.4: Dialogue from the script "Magnet_Err", showing the menu choices of gradient or angular, dipole, error. As indicated above, both error types can be characterized as perturbing "kicks".

the perturbations are added and rays are tracked through eight FODOs. The results for an F quadrupole with a focal length of 500 m added at the centerline of the F quadrupole of the MR FODO, as in Eq. (12.4), appear in Figure 12.5.

Another set of randomly chosen rays appears in Figure 12.6 for the situation where a bend is introduced into the FODO at the Q_F location. The rays repeat after about five FODOs, as is also the case in Figure 12.5 since the lattice betatron results are not modified from the design values.

In this simple example, a single element of a single FODO is perturbed. The focal length of the F quadrupole is altered or the bend angle of a FODO dipole is changed. In a more general case, suppose that there is some imperfection which causes a driving term in the Hill equation characterized as a force, $F(s)$. That term will cause a displaced equilibrium orbit. If Q is an integer, the displacement will diverge since the system is driven at a natural frequency of the system, which implies a resonance since successive revolutions reinforce one another. Many classical systems in mechanics display resonance behavior if a driving term oscillates at a natural frequency, so this is not unexpected. Such resonant behavior has already been observed in Eq. (12.4). If the tune, Q, is an integer, the "kicks" due to quadrupole misalignment or dipole field errors will excite resonance behavior.

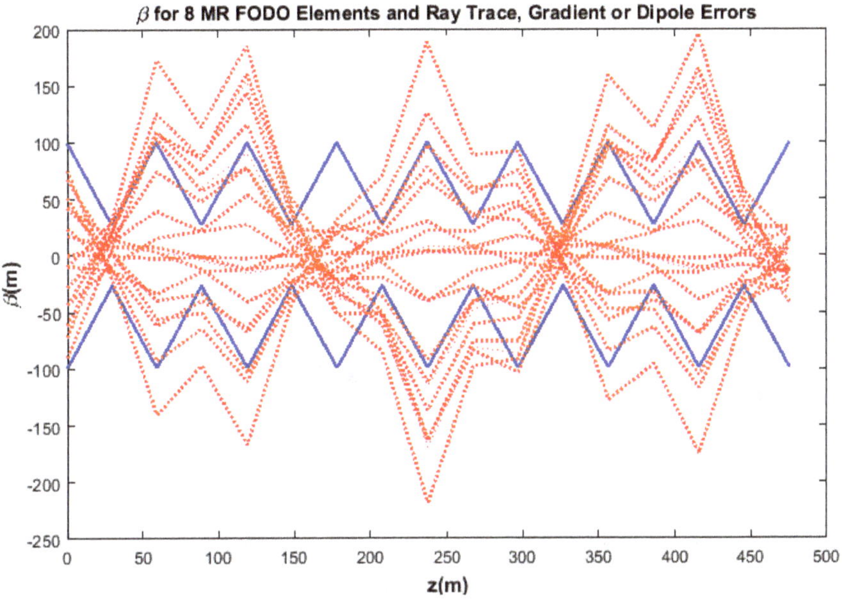

Figure 12.5: Ray traces for an F quadrupole of focal length 500 m added to the normal MR FODO. The stronger focusing causes an overshoot at downstream F quadrupoles. The betatron frequency is unchanged, as expected.

Turning to quadrupole gradient errors, such an error can be thought of as the inclusion of an additional thin lens, as shown in Eq. (12.6), where the error in k is dk:

$$\begin{pmatrix} 1 & 0 \\ -(k+dk)L & 1 \end{pmatrix} \sim \begin{pmatrix} 1 & 0 \\ -kL & 1 \end{pmatrix} \begin{pmatrix} 1 & 0 \\ -dkL & 1 \end{pmatrix}. \tag{12.6}$$

The additional thin lens changes the transfer matrix and thus the tune is changed, as is shown below; the printout of the script "Quad_Grad_Err" appears in Figure 12.7. Generically, the Hill equation in the linear approximation has a driving force $F(s)$ which is proportional to displacement. In that case, the effect of the linear perturbation can be considered to be embodied in a phase shift for the reference orbit, $F(s) = Gx$, $Q^2x \to (Q^2 - G)x$, $\delta Q \sim -G/2Q$. Any force can be expanded about an equilibrium point. The linear term can then be subsumed, if small enough, into a tune shift. Higher order terms cannot be treated in this linear fashion and lead to nonlinear

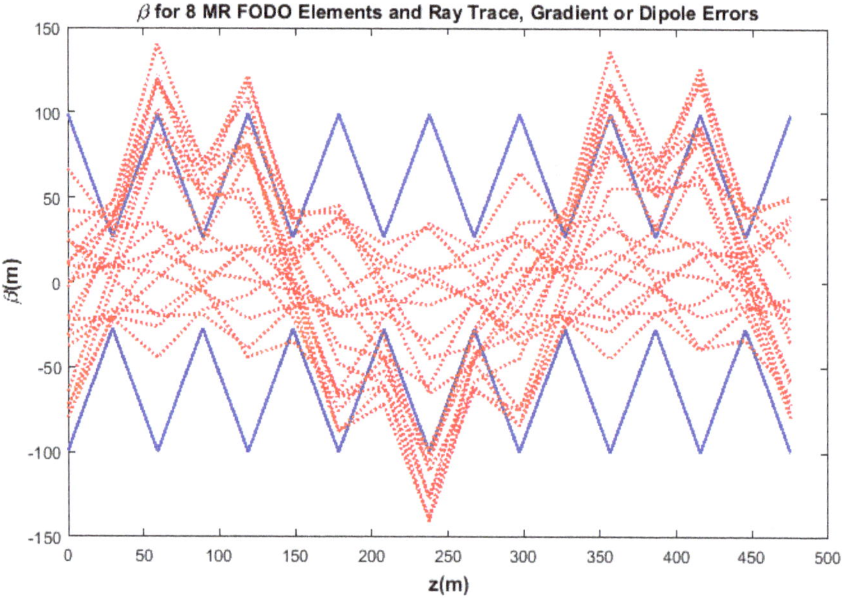

Figure 12.6: Ray traces for an added impulse in angle of 50 mrad in the centerline of the F quadrupole. There is an overshoot at downstream F quadrupoles.

behavior. The emittance will no longer be a conserved quantity and will grow in general. A specific example for particle acceleration will be discussed later in the text.

The "tune shift", using trig identities again, is simply $dQ = \beta dkL/4\pi$. There is a tune shift in both planes. For a horizontally focusing quadrupole, an increase in gradient causes an increase in Q_x but a decrease in Q_y. The shift is $dQ = \beta(dk/k)/4\pi f$ which means that a 1% gradient error in a single MR F quadrupole induces a 0.003 tune shift.

A simple generic example of nonlinear terms in an equation with constant k is explored in the script "Nonlinear_Hill", where the equation is $d^2x/d^2t = x + dx^2$ and the value of d is supplied by the user. The equation is solved over a fixed time range using the utility "ode45". A specific result is shown in Figure 12.8, where the $(x,\, x')$ phase space occupied by the system is plotted for three different values of $d; d = 0.1, 0.3$, and 0.5. The growth of the emittance as the quadratic coefficient grows is quite evident. The oscillation

```
>> Quad_Grad_Err
 Quadrupole Gradient Error

Mg =

[     1, 0]
[ -dkL, 1]

 Modified Transfer Matrix Trace

ans =

cos(w) - (b*dkL*sin(w))/2
```

Figure 12.7: Output of the script to look at the change in the trace of the transfer matrix with the addition of a thin lens representing a small quadrupole gradient error, $M->M*M_g$.

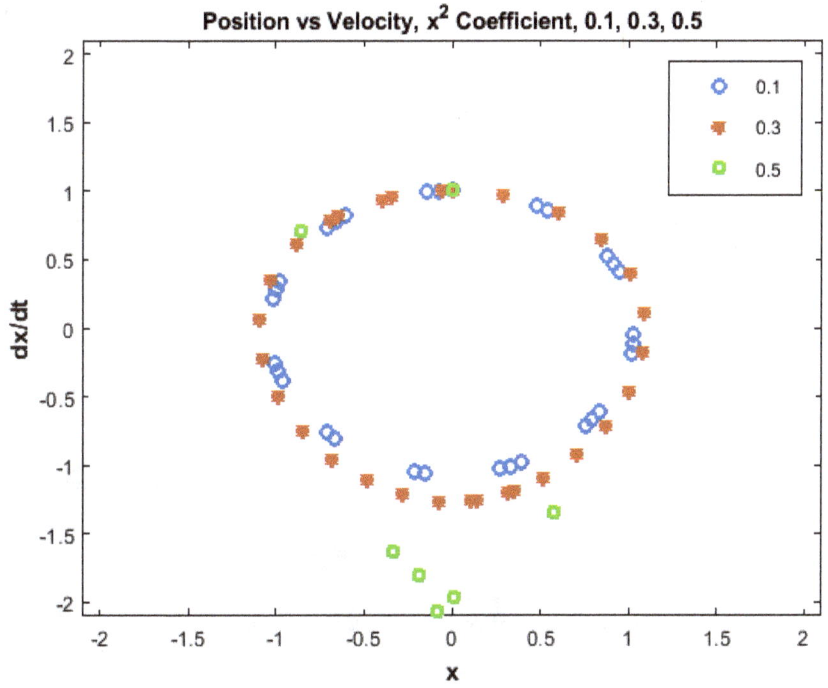

Figure 12.8: Phase space trajectory for three different values of the quadratic coefficient.

frequency also decreases as d is increased, as can be seen by the distance between points at fixed time intervals or by the "movie" supplied by the script. At a d value of 0.5 the system is unstable and constrained motion is no longer possible. These are generic features with nonlinear forces.

The gradient error also causes a shift in the β function. The change is worked out in the script "Quad_Grad_Err_General". The transfer matrix, Eq. (12.2), is used in the simplifying special case where $\alpha = \alpha_o = 0$, or at the location of a quadrupole which is not terribly restrictive. The one-turn matrix $M(s_o + C, s_o)$ with and without the perturbation, Eq. (12.6), is propagated to the location s using the transformation $M(s|s_o)M(s_o + C, s_o)M^{-1}(s|s_o)$ to find $M(s + C, s)$. The printout from the script appears in Figure 12.9.

The fractional change in the β function, after some trig manipulation, is

$$d\beta(s)/\beta(s) = -(dkL)\beta_o[\cos(\omega) - \cos(2\Delta\psi + \omega)]/[2\sin(\omega)]. \tag{12.7}$$

```
>> Quad_Grad_Err_General
 Phase Change dw for Gradient Error

dcosw =

-(bo*dkL*sin(w))/2

 Change in bsinw for Gradient Error

dbsinw =

-bo*dkL*sin(dphi)*(b*sin(dphi)*cos(w) + cos(dphi)*sin(w)*(b*bo)^(1/2)*(b/bo)^(1/2))

>> subs(dbsinw,(b*bo)^(1/2)*(b/bo)^(1/2),b)

ans =

-bo*dkL*sin(dphi)*(b*cos(dphi)*sin(w) + b*sin(dphi)*cos(w))

>> pretty(simplify(ans))
-b bo dkL sin(dphi + w) sin(dphi)
```

Figure 12.9: Printout to evaluate the change in β at an arbitrary location s due to a quadrupole gradient error at location s_o. Command line input is made to clean up the symbolic expressions.

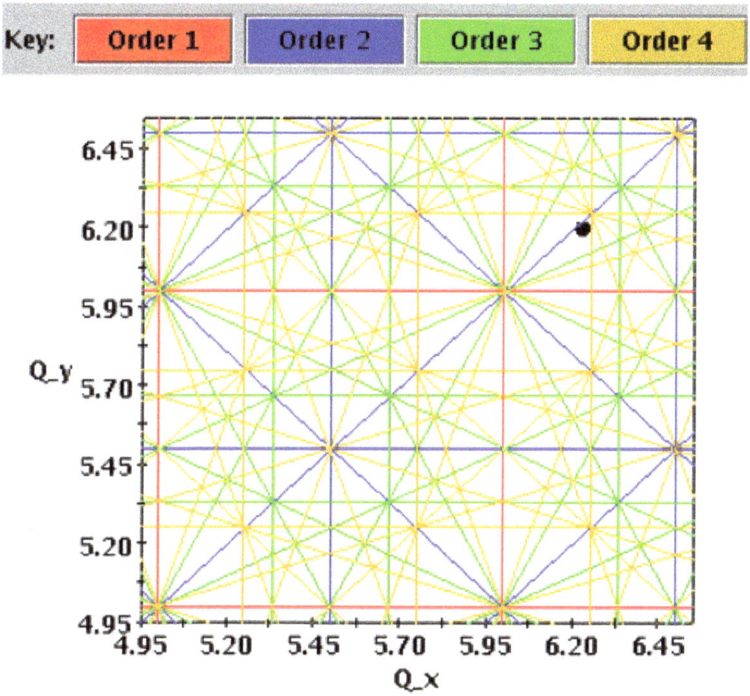

Figure 12.10: Typical tune diagram for horizontal and vertical tunes. Integer resonances at $Q = 5, 6$ are indicated, as well as half integer resonances at $Q = 5.5$, 6.5. Resonances for coupled x and y motion at "order 2" are also indicated, as well as higher order resonances.

In this case there is also resonance behavior, but at half integer values of the tune Q. There is a "beat frequency" which has twice the tune frequency, which means that the disturbance is largest halfway around the ring. A diagram of the x and y tune with integer and half integer resonances to be avoided appears in Figure 12.8. The script also returns the change in the x value. The result for $x'_o = 0$ is $dx/x_o = -(dkL\beta_o/2)[\sin(\omega) + \sin(\omega + 2\Delta\psi)]$.

Using a beam position monitor (BPM) and giving a tweak to a specific quadrupole, the quadrupole gradients can be determined by measuring the change in β. Alternatively, the phase shift can be measured in order to assess the change due to the tweak. A plot of the orbit centroid can be used to track deviations from the ideal closed orbit. Those deviations can, in turn, be corrected for with small

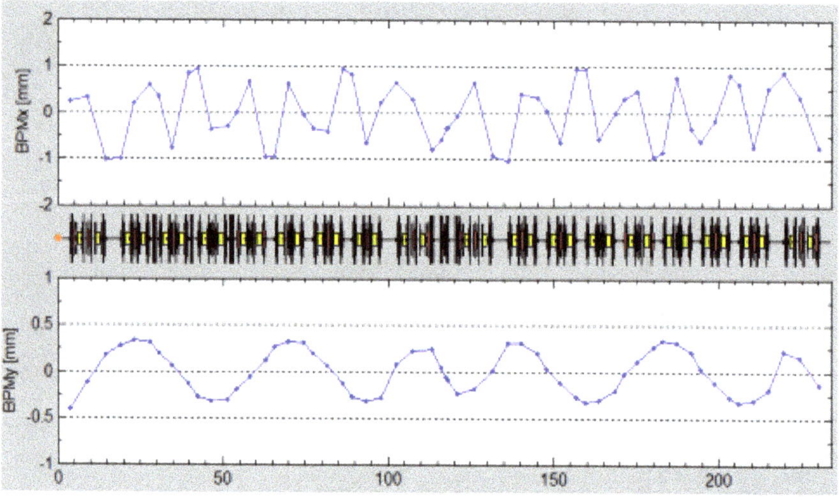

Figure 12.11: Simulated β function data in the presence of a quadrupole error of 0.1 mm in both x and y.

adjustments to the elements or by using distinct weak correction elements. A plot of a simulation of closed orbit errors due to small gradient errors appears in Figure 12.11. A slow "beat frequency" due to the error is evident.

A related effect is due to the off-momentum particles in the beam. A quadrupole has a shorter focal length, $1/f = kL$, for low momentum particles and a longer one for high momentum particles. There will be a tune shift for off-momentum particles, which means that the tune is not a fixed value but is broadened by off-momentum beam particles. The ideal particle has a closed orbit focal length of f_o.

$$\begin{aligned} 1/f &= 1/f_o + d(1/f) \sim 1/(p_o + dp) \\ d(1/f) &= -(dp/p)/f_o \\ dQ &= -(\beta_o/4\pi)(1/f_o)(dp/p). \end{aligned} \tag{12.8}$$

Weaker focusing is favored if the chromatic tune broadening is to be kept small. Formally, this result is the same as the one above for a quadrupole fractional gradient error. A 1% momentum acceptance would lead to a tune spread of about 0.003.

There is both a shift in tune for the particles having the central momentum and a width in tune due to chromatic effects. The fractional shift, $\delta Q/Q$, and the fractional broadening are both approximately proportional to β^2 ([Eq. (12.1), $Q \sim \rho/\beta$]). Low values of β are preferred in order to minimize both tune shift and tune broadening.

In the case of a quadrupole rotation by an angle θ, there is an angular coupling of the x and y motion, $\delta x' \sim 2y\theta/f$, $\delta y' \sim 2x\theta/f$. Such a coupling requires a four-dimensional matrix treatment, which will not be attempted here.

For a dipole rotated in the $(x,\ y)$ plane by an angle θ, an additional vertical angle is induced: $\delta y' \sim \phi_B^2\theta/2$. There will then no longer be a median plane, and the x and y motions will be coupled. A resonant instability occurs when $Q_y + Q_x$ or $Q_y - Q_x$ is an integer, as illustrated by the diagonal lines in Figure 12.10. A rotated dipole creates a "stopband", again with a width due to chromatic effects.

These errors can be mitigated by quality control in manufacturing, in precision survey and alignment, and in the addition of small and weak correction elements. The position and field errors are normally not an issue, at least over an aperture where the fields are near their design values. Therefore, these issues are largely issues that can be alleviated by engineering.

Until now the beam of particles has been assumed to propagate in a perfect vacuum. However, that is not correct and, in addition to the quality of the vacuum, diagnostic equipment may be used to track the beam, which requires a finite amount of material. The encounter of a beam particle with a particle in the vacuum can lead to elastic scattering or an inelastic interaction. The latter will likely lead to loss of that particle, while the former may be recovered if the scattering angle falls within the acceptance ellipse of the accelerator.

A table of the "radiation length", or X_o, of materials appears in Appendix F. The multiple scattering angle was already used in the exploration of a muon beam in Chapter 6. This stochastic process depends on the square root of the amount of material traversed. A multiple scattering RMS angle is determined by the length of material traversed, x, and by the momentum and velocity of the charged beam particle. The use of x is conventional. It would be z if not for that convention.

$$\begin{aligned}\varepsilon_s &= 21.2 \text{ MeV} \\ \theta_o &= (\varepsilon_s/p\beta)\sqrt{x/X_o}.\end{aligned} \tag{12.9}$$

As can be seen from Eq. (12.10), the multiple scattering can be thought of as giving a transverse momentum impulse, $\Delta p_T = (\varepsilon_s/\beta)\sqrt{x/X_o}$ to the beam particle. For high energy beams, since the characteristic scattering energy is ε_s, the scattering angle can be small for thin material crossings and the beam can remain stable if the emittance growth remains within the accelerator acceptance (limiting apertures). The vacuum issues are most stringent in storage rings, where the beams must be stable for several hours. Indeed, the LHC vacuum is of better quality than that found on the Moon for just this reason.

There are issues which are more intrinsic to the formation of intense beams and which cannot so easily be mitigated because the greatest fundamental difficulty arises from the charge of the beam particles themselves. As the luminosity of colliders increases and as the intensity of fixed target beams increases, — "the intensity frontier", — these physics issues gain more importance.

Thus far the beam itself has been considered to be an ensemble of noninteracting particles guided solely by external electric and magnetic fields. However, the particles in a beam also repel one another and are therefore subject to additional forces. Indeed, at some point these forces may seriously perturb the orbits defined by the external forces and lead to beam loss. There is a limit to the particle density that can be contained stably by external forces since mutual Coulomb repulsion of the beam particles grows with beam density.

Consider a beam as a cylinder of moving charge, assumed to be continuous and not an ensemble of point particles. Because of symmetry there is only a radial electric force and an azimuthal force due to the magnetic field caused by the beam current. The coordinates are shown in Figure 12.12. The Lorentz force, Eq. (12.11),

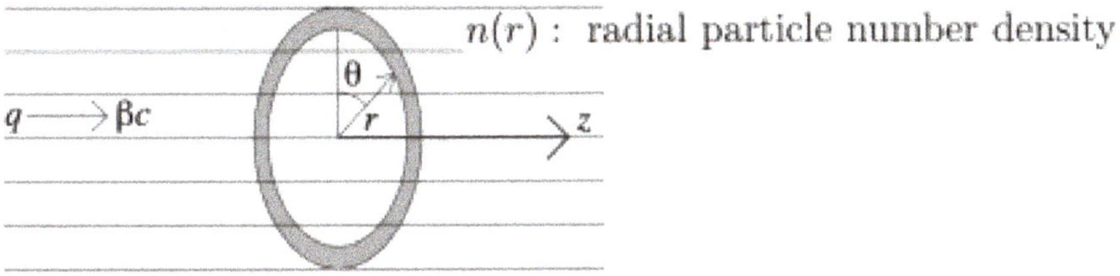

Figure 12.12: Coordinates of the beam charge density in the laboratory frame.

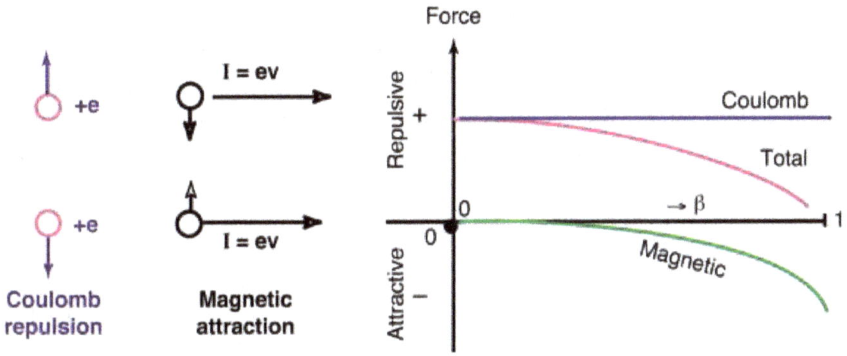

Figure 12.13: Visualization of the electric and magnetic forces on two particles in the beam. Like charges repel but coaxial currents attract.

is purely radial.

$$F_r = qe(E_r - vB_\theta). \tag{12.10}$$

The electric and magnetic forces can be visualized as shown in Figure 12.13. There is Coulomb repulsion of any two charges. However, the moving charges are currents and like currents attract due to their magnetic fields. As the velocity approaches c, the net force approaches zero.

For a radial number density, $n(r)$, Maxwell's equations for the divergence of E and the curl of B in terms of the charge and current are

$$\begin{aligned} \vec{\nabla} \cdot E_r &= 1/r[d/dr(rE_r)] = qen(r)/\varepsilon_o \\ (\vec{\nabla} x \vec{B})_z &= 1/r[(d/dr(rB_\theta)] = qe\beta n(r)/c\varepsilon_o. \end{aligned} \tag{12.11}$$

The divergence and curl equations are easily integrated to solve for the fields.

$$E_r = (qe/\varepsilon_o r)\int n(r)rdr, \quad B_\theta = (\beta qe/c\varepsilon_o r)\int n(r)rdr = (\beta/c)E_r. \tag{12.12}$$

The total force is then nearly zero for ultrarelativistic particles. The effects of the "space charge" are therefore most important at low energies, specifically at injection. For a constant radial density, the electric field is proportional to n and r and vanishes at r of zero by symmetry. The overall force has a factor $1 - \beta^2 = 1/\gamma^2$, which tracks the cancelation of the total force due to the magnetic field.

For a more realistic case with a Gaussian beam with RMS, σ, transverse size, bunch length L and with N total particles in the bunch,

$$\begin{aligned} n(r) &= (N/L)(1/2\pi\sigma^2)e^{-r^2/2\sigma^2} \\ F(r) &= (Ne^2q^2/2\pi\varepsilon_o L)(1/\gamma^2 r)(1 - e^{-r^2/2\sigma^2}). \end{aligned} \tag{12.13}$$

The force is defocusing, as expected, since it is repulsive, and it is linear in r for r values less than σ. A plot made by the script "Charged_Beam" given in Figure 12.14 shows the linear behavior at small r on a scale set by σ and the highly nonlinear shape at large r. The linear part of the force can be treated as a tune shift, as has been done previously with the linearized magnet errors.

This force results in a modified Hills equation with an inhomogenous term analogous to a driven harmonic oscillator. Transforming from r to x, the force in the x direction is proportional to x. Converting to dt from ds, $ds = vdt$, the modified force/acceleration equation becomes, using $s = \rho\theta$, with mean radius ρ to shift from s to the angular phase,

$$\begin{aligned} &F = dp/dt = \gamma m dv/dt = \gamma m v dv/ds = \gamma m v^2 d^2x/d^2s \\ &F(x) = (Nmc^2 r_o/L\gamma^2\sigma^2)x \\ &d^2x/d^2s = F(x)/m\gamma v^2 = x[(N/L)r_o]/(\sigma^2\gamma^3\beta^2) \\ &d^2x/d^2\theta + Q^2x, \quad s = \rho\theta. \end{aligned} \tag{12.14}$$

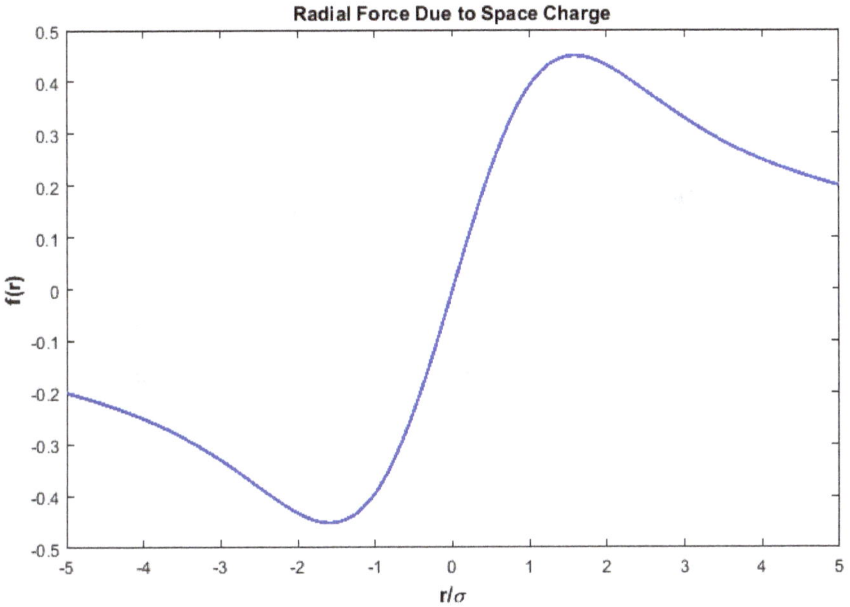

Figure 12.14: Radial force on a beam with a Gaussian radial density distribution. The force is approximately linear for $r < \sigma$.

The velocity factors arise from the t-to-s conversion, while the γ factor arises from going from force to time rate change of momentum. There is a force term linear in x for small deviations, which leads to harmonic motion. The charge is assumed to have $q = 1$. The classical radius of the beam particle is r_o, a distance which occurs when the electromagnetic self-energy of the particle is equal to the mass: $e^2/4\pi\varepsilon_o r_o = mc^2$.

The linear part of $F(y)$ causes a simple tune shift, just as do other linear perturbations. Assuming a tune shift $Q \to Q + \delta Q_{sc}$, the linear part of the space charge tune shift is $\delta Q_{sc} = -(N/L\sigma^2)(r_o\rho^2)/(2Q\beta^2\gamma^3)$. The shift is proportional to the number per unit length within the bunch and falls off rapidly with beam energy.

For protons the classical radius is 1.5×10^{-18} m. For accelerators such as the Fermilab Booster, with only 200 MeV injection kinetic energy from the Linac, the tune shift is large and not exciting a vertical resonance, $\delta Q_{sc} < 1$, where $Q = 6.7$ for the Booster, limits the bunch intensity N to a value which is about 6×10^{10} protons per bunch. With an emittance of $12\pi^* m^*$mrad and a lattice $\beta \sim 33$ m with kinematic variables $\beta\gamma \sim 1$ at injection, the value of σ is about 8 mm. For a bunch length of 1 m and a Booster radius of 75 m, the tune shift is about 0.39. Note that the Fermilab Linac has since been upgraded with an increased output energy to 400 MeV, in order to reduce the space charge tune shift at injection into the Booster.

To illustrate the limits, a numerical example is plotted in the script "Space_Charge". The defined parameters are $R = 50$ m, $Q = 6$, $\beta = 0.3$ (relativity, not betatron), and $L = 1$ m. The tune shift δQ as a function of the number in the bunch, N, and the transverse size of the bunch, σ, is plotted in Figure 12.15. The plot is logarithmic because of the power law behavior of the tune shift on the variables. The rapid decrease in the tune shift with increasing transverse beam size and a decreasing number of particles in the bunch is quite

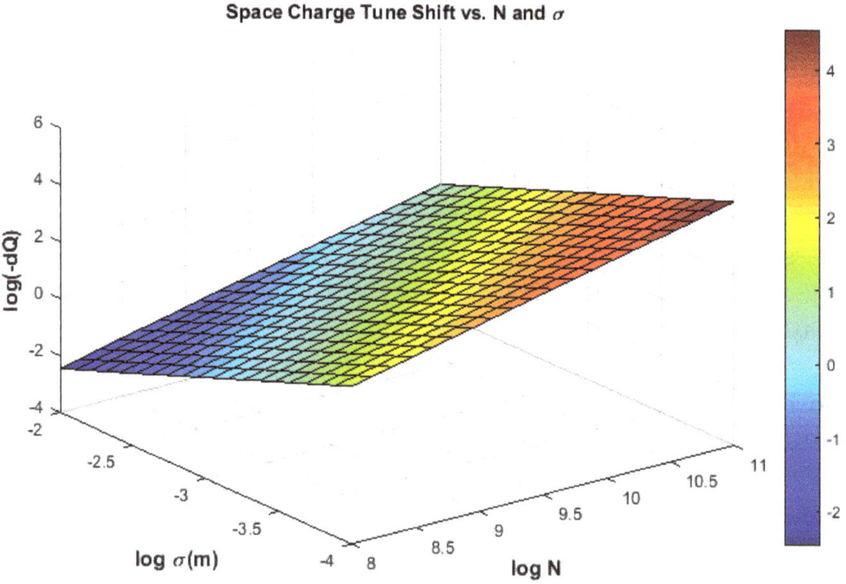

Figure 12.15: Plot of $-\delta Q$ as a function of N and σ. For the values of the other fixed parameters, the tune shift exceeds 1 for large N and small σ.

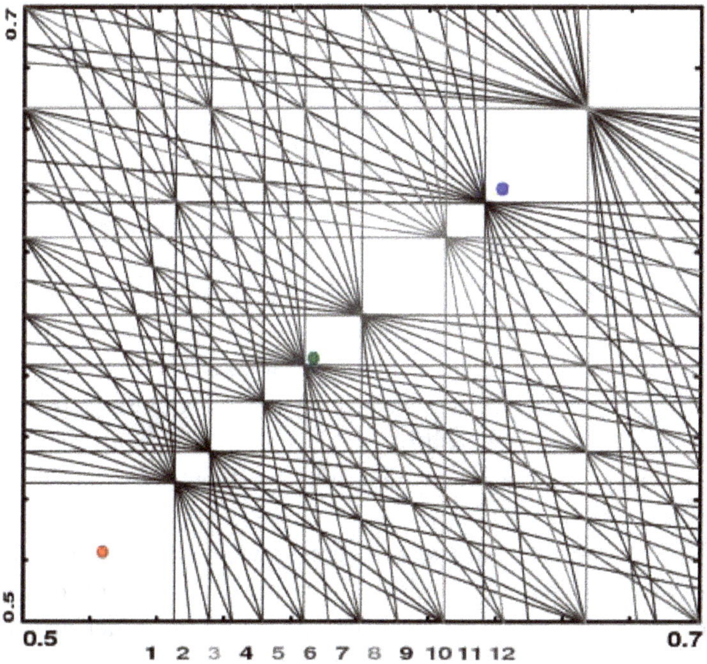

Figure 12.16: Plot of Q_x and Q_y for the Fermilab Tevatron collider, showing operating points placed away from various resonances. The integer tune is 19 with small additions shown here. Tune shifts and spreads smear out these points and are guarded against.

evident. The user can easily edit the script to change any of the fixed parameters. It is expected that intense low energy accelerators will have large space charge tune shifts.

A plot of some of the resonant lines and the operating points in Q_x, Q_y space is shown for the Tevatron Collider in Figure 12.16. The operating "points" are smeared out by various tune shifts and spreads, such as gradient errors and space charge.

The tune shift refers to the transverse betatron oscillations and their perturbations. In addition, there are repulsive forces along the z direction which affect the bunch structure. For a uniform beam of radius a and a conducting beam pipe aperture of radius b, in the CM of the beam (beam at rest) this becomes a standard electrostatic problem of a charged rod in a conducting cylinder.

$V_{CM} = (qe/4\pi\varepsilon_o)\{(N/L\gamma)[1 + 2\ln(b/a)]\}$, where the γ factor arises from the length contraction. For a more realistic parabolic falloff of the bunch at the edges, the laboratory force along the beam, z, direction is derived from the gradient of the potential $V(z) = (qe/2\pi\varepsilon_o)\{(3N)(z/L\gamma)^2[1 + 2\ln(b/a)]\}$:

$$F(z) = (3Ne^2q^2/\pi\varepsilon_o)\{[(z/L^2\gamma^2)[1 + 2\ln(b/a)]\}. \tag{12.15}$$

The z coordinate is measured from the peak of the bunch density. This longitudinal force will drive a shift in the synchrotron frequency, not the betatron tune, which will spread out the bunch in z. The bunch then contains particles oscillating at different frequencies dependent on their z position. This effect is important near the crucial transition energy, which will be mentioned later, in the chapter on acceleration.

The circulating beam also interacts with the conducting walls of the vacuum chamber. Ignoring the magnets and other devices, assume that the beam is a line charge with linear density λ. In a simple one-dimensional case, the beam is centered on a vacuum wall of half height h. A schematic layout appears in Figure 12.17.

In order to meet the wall boundary conditions, image charges must be introduced to make the parallel electric field component zero. The first pair of image charges cancels the effect of the beam charge but then must themselves be canceled.

A line charge at a distance d from the observation point has a field $E = \lambda/2\pi\varepsilon_o d$. An infinite series of image charges is needed to sum the total parallel field at the wall to be zero.

$$\begin{aligned} E_y^n &= (-1)^n\lambda/(2\pi\varepsilon_o)[1/(y + 2nh) - 1/(-y + 2nh)] \\ &= (-1)^n\lambda/(4\pi\varepsilon_o)[y/(nh)^2] \\ E_y &= \lambda/(4\pi\varepsilon_o)[\pi^2/12h^2]y. \end{aligned} \tag{12.16}$$

A line charge at a distance d from the observation point has a field $E = \lambda/2\pi\varepsilon_o d$. An infinite series of image charges is needed to sum the total parallel field at the wall to be zero. The series can easily be summed using the MATLAB utility "symsum", as shown in Figure 12.18.

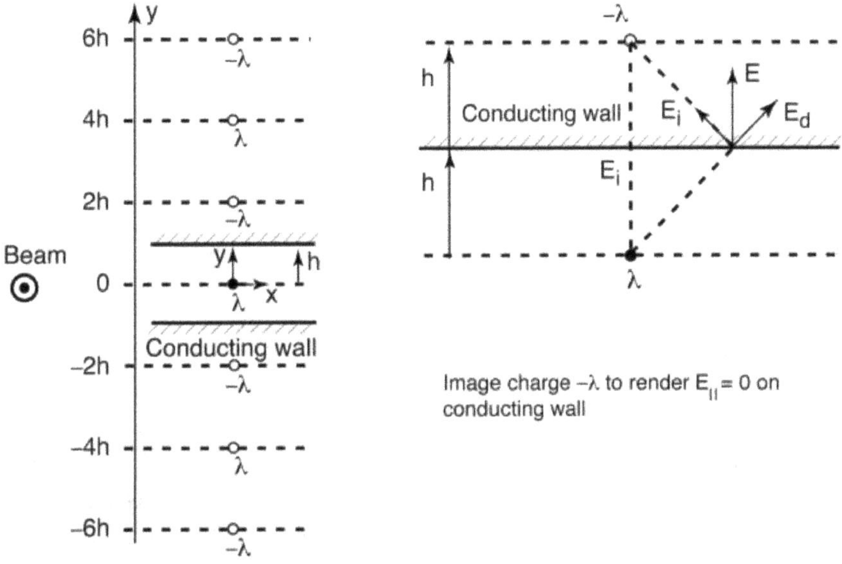

Figure 12.17: Beam as a line charge confined between two walls with total separation $2h$, showing the required image charges which satisfy the boundary conditions.

```
>> S = symsum((-1)^k/k^2,k,1,Inf)

S =

-pi^2/12
```

Figure 12.18: Use of the utility "symsum" to sum the image charge series.

An interesting application of these image charges in the vacuum chamber is using them to construct a BPM. A schematic of such a device appears in Figure 12.19. Any difference between the induced charges left and right indicates that the beam is not centered on the vacuum pipe.

Both the x and y fields in the full three-dimensional case can be found using the Maxwell equation $\partial E_x/\partial x = -\partial E_y/\partial y$ so that E_x is proportional to x and is negative. The forces due to those field lead to both x and y tune shifts, the x shift being positive

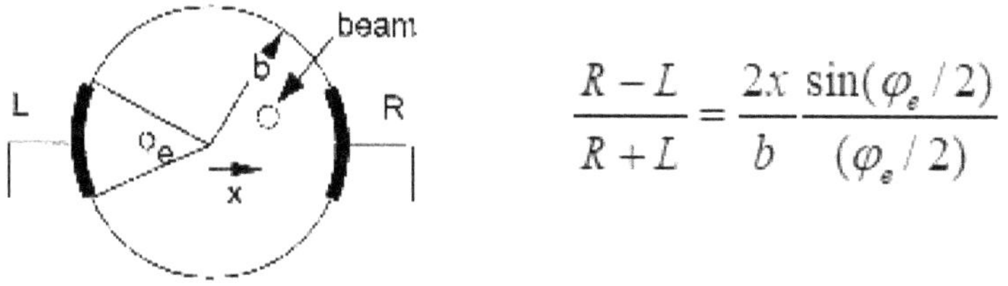

Figure 12.19: Schematic diagram of a BPM created by using the induced image charges in the vacuum chamber.

and the y shift negative. They are similar to the space charge tune shift, with the substitution $(1/\sigma\gamma)^2 \to (\pi/h)^2/12$. The lack of the γ factor is because this is only an electrostatic problem. The effect is focusing in x and defocusing in y and increases with decreasing h. Relative to space charge it becomes more important as the beam energy increases. For example, for $\sigma = 20\mu$m and $h = 1$ cm, the two forces are equal at a factor of $\gamma = 550$, which is not a particularly high energy beam.

Beam–beam forces are also important in colliders and they introduce limitations on both the total luminosity of colliders and the low beta insertions. The magnetic forces for the currents change sign with respect to the space charge, because the beam charges are now of the same sign but opposite direction for a proton–proton collider. From the point of view of Figure 12.10, unlike currents now repel. In that case the $1/\gamma^2$ factor becomes $1 + \beta^2 \sim 2$ and the effect is no longer confined to low energy situations. As before, the force in the horizontal direction can be integrated over the bunch length, L, and an impulse approximation can be made for the overall effect. The linear term for the force in x, Eq. (12.15), becomes

$$F(x) \sim (e^2/2\pi\varepsilon_o)(N/L\sigma^2)x. \tag{12.17}$$

There are N protons in a bunch of z length L and transverse RMS size σ. Again, there is a linear force so that the solution can be cast in the form of an induced tune shift. The resulting momentum impulse due to the beam–beam force is $\sim F(x)(L/2c) = dp_x = Npr_o/\gamma\sigma^2$. This thin lens quadrupole due to the crossing of the two colliding bunches in the (x, z) plane of the reference orbit makes

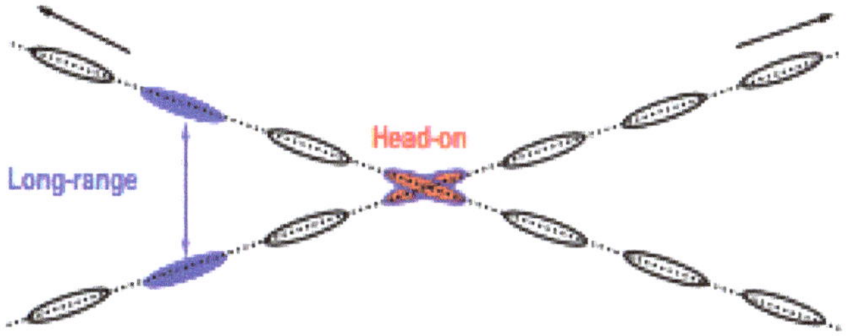

Figure 12.20: Colliding proton beams at the LHC with 25 nsec bunch spacing. Avoiding nonlinearities in the tune shift argues for large separation of the beams, but achieving maximum luminosity argues for head-on collisions.

an angular disturbance $\delta x' = (N r_o/\gamma\sigma^2)x$ which is effectively a lens which causes a change in the overall focal length which results in a beam–beam tune shift. The previous result for a quadrupole gradient error, $dQ = \beta dkL/4\pi$, can be applied to the beam–beam case; $\delta Q_{bb} \sim \beta_x^*(\delta x'/x)/4\pi = (\beta_x^*/4\pi)(N r_o/\sigma^2\gamma)$ for each interaction point in the accelerator.

The β value in this instance is the specific β^* value at the interaction point, which is small rather than an average $\langle\beta\rangle$ around the ring. The tune shift is localized near the interaction point and there is no phase averaging around the ring. This means that the beam–beam tune shifts must be tightly controlled to a typical value of about 0.01 or less. Using the LHC as an example, $\sigma = 17\mu$m, $N = 1.1 \times 10^{11}$ p/bunch, and $\beta* = 0.5$ m. The linear tune shift can be of order 0.035, but the long range nonlinear effects can be more troublesome since the whole other beam contributes.

This force is nonlinear for large x, as mentioned before. The resulting nonlinearities argue for going to large separation of the beams. However, high luminosities argue for head-on collisions, or at least small crossing angles. A schematic picture of the colliding beams at the LHC appears in Figure 12.20. These competing effects are in tension, and at the LHC are planned to be mitigated by "crab cavities" which rotate the beams from a large crossing angle to head-on collisions quite near to the interaction point. The beam–beam tune shift limitations are then localized to the interaction regions. The total tune shift is proportional to the number of interaction regions, which makes running multiple experiments more difficult.

Chapter 13

Acceleration

We're like children who always want to take apart watches to see how they work.

— **Ernest Rutherford**

Before looking into the longitudinal acceleration of a particle beam, an analogous situation arises in the case of the mechanics of a pendulum executing very large angle motion. The dialogue for the script "Large_Angle_Pendulum" appears in Figure 13.1.

The script uses the MATLAB utility "ode45" to solve the appropriate differential equation numerically. A plot of the small angle, simple harmonic motion and the large angle angular velocity appears in Figure 13.2. In the large angle case the oscillation is clearly not sinusoidal.

A plot of the angular displacement and the angular velocity appears in Figure 13.3. This is a way to present the data which removes time as an independent variable and is a common way to present the phase space contour of a system. For small angles the path of sinusoidal oscillations appears to be circular. In contradistinction to that behavior, for large angles, the path of the pendulum extends to angular values of π and the path is not circular. A similar behavior will appear when the acceleration of a particle beam is studied. This classical analogue serves to indicate the unity of all physics. The user can generate unbound motion by trying more extreme values for the initial conditions than those that appear in Figure 13.1. This behavior too will appear in considering the acceleration of charged particle beams.

```
>> Large_Angle_Pendulum
Pendulum - Large Oscillations
Enter Initial Angle (degrees): 130
Enter Initial Angular Velocity (degrees/sec): 30
Enter g/L in MKS units: 1
Small Angle Circular Frequency = 1 1/sec
Small Angle Period (sec) = 6.28319
Period = 2 * pi *sqrt(L/g) Increased at Large Angles
```

Figure 13.1: Dialogue used to explore the motion of a simple free pendulum executing large angle motion.

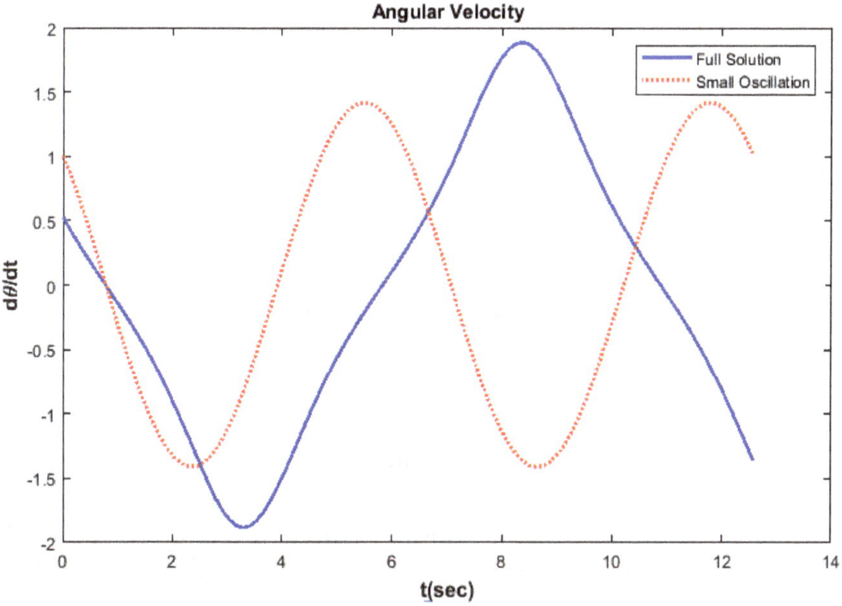

Figure 13.2: Angular velocity of a pendulum for small, harmonic oscillations and for large amplitude excursions.

First, a source of beam particles is needed to serve as input to an accelerator. For protons, the simplest incarnation is to simply ionize hydrogen gas in a strong electric field which disassociates the gas. At Fermilab there was a series of accelerating steps. First, in the Cockcroft–Walton, a voltage multiplier device, the ions gained

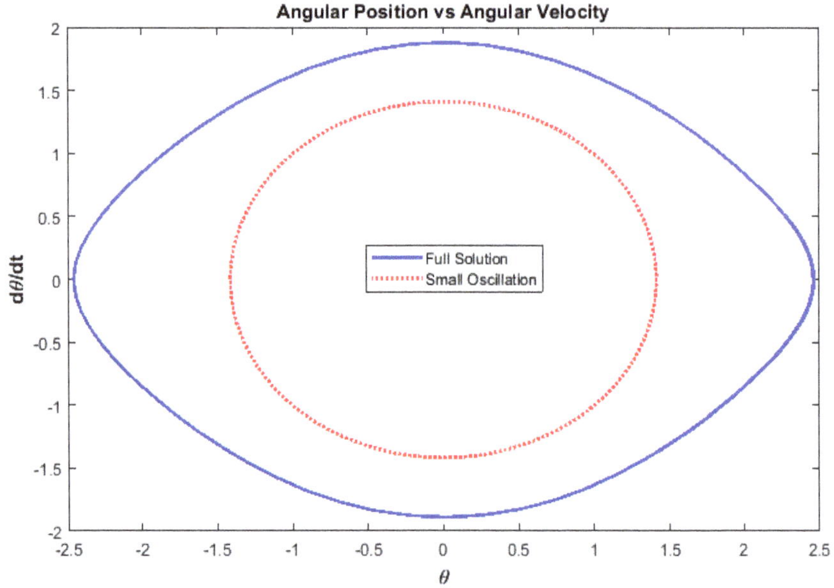

Figure 13.3: Contours of angular position and velocity for small amplitude and large amplitude oscillations.

a kinetic energy of 0.75 MeV. A picture of the Cockcroft–Walton is shown in Figure 13.4. This simple method of obtaining free protons and accelerating them has been superseded by more efficient methods.

The protons were then accelerated in a drift tube linear accelerator which was about 150 m long, where they were acted on by harmonic electric fields which raised the kinetic energy to 200 MeV. The maximum emittance is 6π mm-mrad, and the maximum possible accepted momentum spread is 0.3%. A photograph of the Linac (linear accelerator) appears in Figure 13.5.

The "efficiency" of a linear acceleration section depends on the length. Since the electric field is oscillatory with a frequency called ω_o, a very long section will give no net acceleration as the average value of a sine wave over many periods is zero. If the voltage were constant, a length L of field would give an energy gain of qeV_o, $V_o = E_oL$. Taking the RF phase into account, the constant field energy gain is reduced by a factor, $\sin(\psi)/\psi$, $\psi = \omega_oL/2\beta c$.

Figure 13.4: Photograph of the Fermilab Cockcroft–Walton preaccelerator.

Figure 13.5: Photo of the Fermilab drift tube Linac. The power leads and vacuum chambers are not shown.

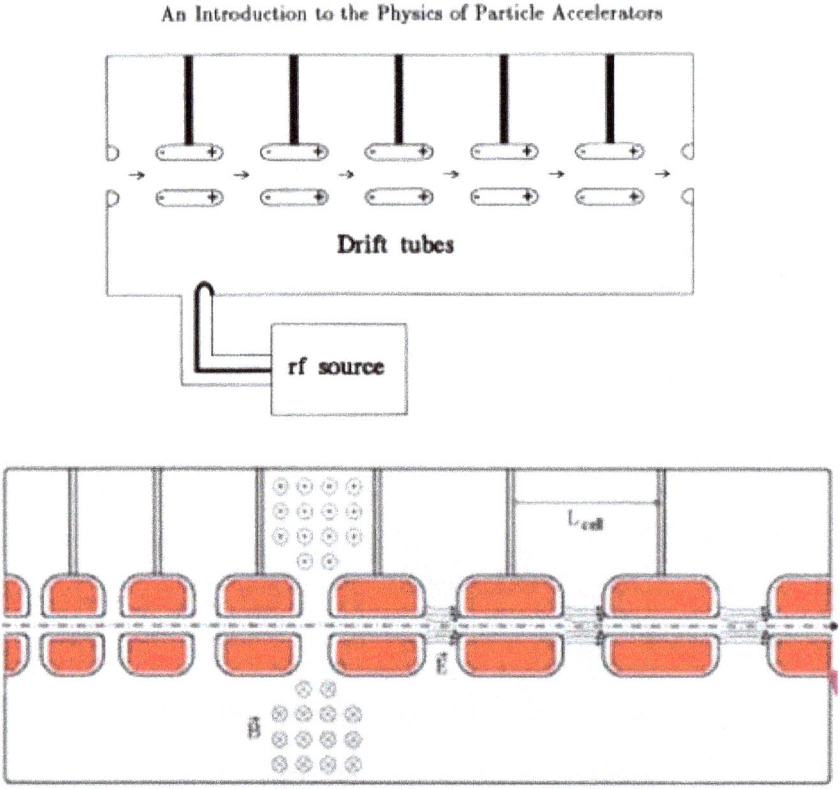

Figure 13.6: Schematic view of the electric fields and locations of the accelerating structures in a linac. *Top*: A very schematic view of the electric fields. *Bottom*: A specific layout (Figure 13.5) showing the electric and magnetic fields.

A very schematic view of the electric fields appears in Figure 13.6. The RF electric fields are longitudinal and the phase location of the different drift tubes is tuned to provide overall acceleration. As the particles gain energy, the spacing of the cavities increases in order to preserve the RF phase relationship.

The linac concept becomes quite long, so the beam is subsequently injected into a circular accelerator, called the Booster at Fermilab, which raises the energy to 8 GeV. When the conventional iron magnets in the Booster reached saturation, the Booster beam was injected into the Main Ring, radius about 1 km, where it was accelerated to 400 GeV. With the advent of stronger superconducting dipoles placed in the Main Ring tunnel, a maximum energy of 1000 GeV or 1 TeV could be achieved. Hence the name "Tevatron".

It is very difficult to achieve a large electric field with a constant field. Therefore, oscillating fields are normally used to accelerate beams. However, only half of the cycle can be used since the other half decelerates. Indeed, the full positive cycle cannot be used either.

A simple metallic cylindrical waveguide is shown in Figure 13.7. The electric field is sinusoidal and longitudinal: $E_z = E(r)e^{i\omega t}$. There must be an accompanying changing magnetic field and it is directed azimuthally. The boundary conditions on the fields are that E must vanish in the conductor wall and B must be perpendicular to the walls. Maxwell's equations for E_z then yield the Bessel equation when cylindrical coordinates are used, as is reasonable given the symmetry of the system. The solutions are $E(r) = E_o J_o(\omega r/c)$, $B(r) = i(E_o/c)J_1(\omega r/c)$. The solution is easily plotted with a few command line inputs, using the utility script "besselj", as seen in Figure 13.8, and where the solutions are plotted in Figure 13.9.

The cavity will only allow an RF which satisfies the boundary conditions. In this case, $\omega_o r/c = 2.405$ for the first zero, as observed

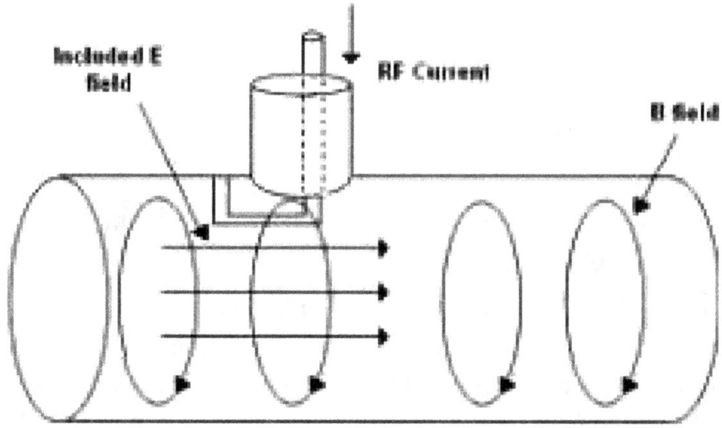

Figure 13.7: Schematic of an accelerating cavity waveguide, showing the electric and magnetic field directions.

```
>> r = linspace(0,2.405);
>> Jo = besselj(0,r);
>> J1 = besselj(1,r);
>> plot(r,Jo,'-b',r,J1,':r',[0, 2.405],[0,0],'-k')
>> title('Jo and J1')
>> xlabel('r')
>> ylabel('Jo, J1')
>> legend('Jo','J1')
```

Figure 13.8: Command line inputs to compute and plot J_o and J_1.

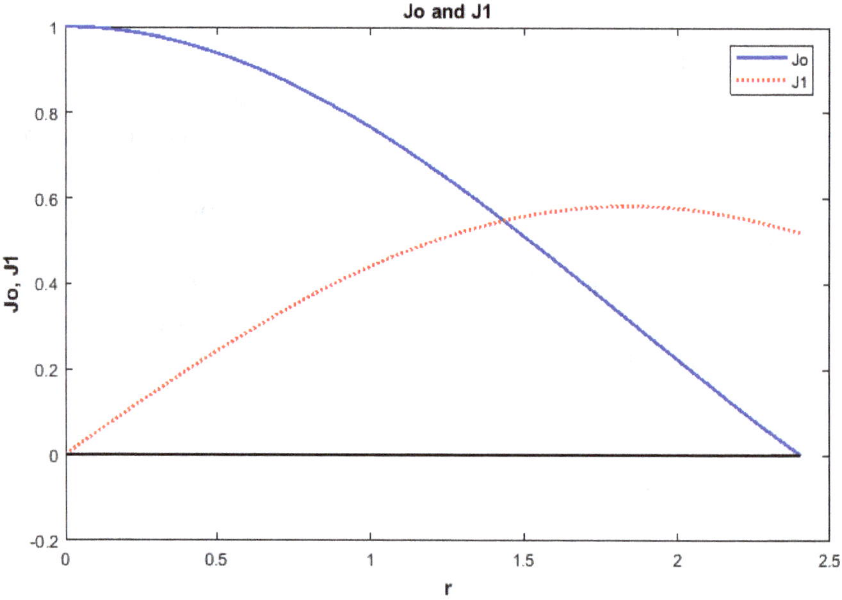

Figure 13.9: Plot of J_o and J_1 as a function of the argument.

in Figure 13.9. The total electromagnetic energy stored in the cavity goes as $J_1^2(2.405)$. The field should be approximately constant over the transverse size of the beam in order that all particles in the beam at a given time in the RF cycle see the same acceleration. That requirement means that the beam size is small with respect to the waveguide aperture, $a = 2.405c/\omega_o$.

Parenthetically, the waveguide has $\partial_z E_z \neq 0$. Since the Maxwell equation, $\vec{\nabla} \cdot \vec{E} = 0$, holds, there must also be transverse fields.

There are two definitions of the emittance of a beam. The amplitude scales as $1/\sqrt{p}$ in the lattice, as has been mentioned previously. The reason is that as the beam is accelerated, the longitudinal component of momentum is increased by E_z, while the transverse components are unchanged. Thus, the angular divergences are reduced. Therefore, at low injection energy, the beam is large and it compresses by "adiabatic damping".

The emittance, or the area of the lattice ellipse, is not invariant under acceleration of the beam particles. As the particles gain energy, the emittance decreases since $p = m\beta\gamma$ increases. Rather, a normalized emittance ε_N is defined which is invariant and accounts for the compression of the angular or x' phase space, $\varepsilon_N = \varepsilon\beta\gamma$. The symbol ε has been used for both energy and emittance, which is unfortunate but which should be clear in the context of the specific discussion.

For a numerical example, consider the Fermilab Main Ring. Beam is injected from the Booster at 8 GeV, $\beta\gamma = 9.5$, and extracted at 400 GeV, $\beta\gamma = 427$. The square root of the emittance has shrunk by a factor of 6.8 between injection and extraction owing to adiabatic damping. At injection the beam RMS transverse size is about 4.6 mm in x and 2.4 mm in y, while at extraction the values are 0.68 and 0.35 mm, respectively. The angles in x and y have also shrunk from 46.6 to 6.9 μrad in x, while the vertical angles evolve from an RMS of 89 to 13 μrad. It is clear that aperture problems are most important at injection.

The Fermilab Linac and all subsequent accelerating devices "bunch" the beams, because only a portion of the phase of the sinusoidal electric field supports stable acceleration. That portion shrinks as the beam is accelerated, so that the bunch narrows longitudinally. In the following a very simple set of assumptions is made. Assume a single accelerating gap per revolution in a circular ring. As the energy increases, the RF must increase so that as the beam speeds up it continues to see a stable phase of the accelerating voltage on each turn.

$$\begin{aligned} \omega &= 2\pi/\tau = 2\pi\beta c/C = \beta c/R \\ \omega_o &= h\omega. \end{aligned} \tag{13.1}$$

The rotation frequency is ω, period τ. The particle velocity is βc. The ring circumference is C, with radius R, assuming a very small accelerating gap. The rest of the ring is filled with ramping dipoles, with the magnetic field increasing to keep the radius of curvature $\rho = R$. The RF voltage has a frequency ω_o which is h times the revolution frequency ω, where h is called the harmonic number. Integer h insures synchronous acceleration at each turn.

For example, at the LHC the revolution frequency at top energy is 11 kHz, while the harmonic number is 35,640 or an RF frequency of 392 MHz. The Main Ring harmonic number is 1113. A particle which sees a fixed stable phase at each revolution is synchronous with rotation frequency ω_s and phase ϕ_s. Other particles in the beam may perform "synchrotron" oscillations about this stable phase and maintain a stable but not synchronous orbit. These are longitudinal oscillation, while the betatron oscillations are transverse. The ideal synchronous particle does not move longitudinally and has a fixed RF phase in analogy with a particle on the betatron reference orbit which stays on that orbit and does not oscillate transversely. Particles with different phases mix longitudinally.

The accelerating voltage is $V(t) = V_o \sin(\omega_o t + \phi_s)$ and time $= 0$ is defined to be when a synchronous particle passes the accelerating gap. The energy gain per turn for the synchronous particle is $qeV_o \sin(\phi_s) = d\varepsilon_s$. It will be shown that there are several phases, with associated z positions, near the synchronous phase which experience a restoring force and which form a "bunch" with "phase stability". However, the phase of particles in the bunch changes with time.

There are two competing effects for momentum which deviates from the synchronous value. First, higher momentum particles go faster and arrive at the gap earlier. However, as β approaches 1 the changes get smaller: $\gamma = 1/\sqrt{1-(1-\delta\beta)^2}$, $\delta\beta \sim 1/2\gamma^2$.

For example, at the Fermilab Booster, the RF goes from 37 to 53 MHz as the energy goes from 0.4 GeV to 8 GeV. Second, higher momentum particles have a larger radius of curvature and travel further in the dipoles. Owing to this dispersion they arrive at the gap later. Linear accelerators have no dipoles and therefore no dispersion. Dispersion means that there is a relationship between dC/C and dp/p, which defines the "transition energy". The value of the transition energy depends on the details of the dispersion in the lattice. For Fermilab, the Main Ring has a value of 18.7 for the transition γ_t factor.

$$1/\gamma_t^2 = dC/C/dp/p. \tag{13.2}$$

The overall effect on the rotation period due to the combined effects is

$$\begin{aligned} \tau &= C/\beta c \\ d\tau/\tau &= dC/C - d\beta/\beta = (1/\gamma_t^2 - 1/\gamma^2)(dp/p) \\ &= \eta(dp/p). \end{aligned} \tag{13.3}$$

For the last step, MATLAB symbolic math makes the calculation relating $d\beta/\beta$ to dp/p easy. The command line dialogue appears in Figure 13.10, where the expression for β is explicitly differentiated using the utility "diff" and $\beta = p/\varepsilon$. The result appears as $d\beta/\beta = (1/\gamma^2)(dp/p)$. Other useful relativistic formulae can easily be derived using MATLAB to avoid some calculus and algebra. For example, $d\gamma = \beta\gamma^3 d\beta$, $d\beta = (\beta/\gamma^2)(dp/p)$, $(dp/p) = (1/\beta^2)(d\varepsilon/\varepsilon)$. In Eq. (13.3) the symbol η is related to dispersion but it is not the dispersion which was defined previously. This is the accepted notation and should not cause confusion.

Below the transition energy the bunching is stable on the rising phase of the RF oscillation. The parameter η is <0 below transition energy and >0 above it. If the beam energy increases and crosses the transition energy, the RF needs to rapidly change phase. A schematic of the changes coming at transition is given in Figure 13.11. The process of going through transition is complex and will not be pursued further here. In either case there is a restoring force which

```
>> syms p m b
>> db = diff(p/sqrt(p^2+m^2));
>> dbb = db/(p/sqrt(p^2+m^2));
>> simplify(dbb)

ans =

m^2/(p*(m^2 + p^2))

>> pretty(ans)
      2
     m
-----------
    2    2
p (m  + p )
```

Figure 13.10: Command line inputs to symbolically compute $d\beta/\beta$ in terms of p and dp/p.

is the cause of the harmonic motion of particles in the bunch that are near to the synchronous orbit location in phase.

For the Main Ring, the maximum RF voltage per gap is about 2 MV. The maximum beta function is about 100, while the maximum dispersion is about 6 m, which means a 6 cm aperture for a 1% momentum spread. The dipole bend angle per full FODO cell is about 66 mrad.

Ignoring the issues with transition from now on, the restoring force and phase stability are examined. For one gap per turn the particle energy and the phase seen by the particle both change in going through the gap. For the synchronous and asynchronous particles the energy gain is $d\varepsilon_s = qeV_o \sin\phi_s$, $d\varepsilon = qeV_o \sin\phi$, respectively. The synchronous energy gain is subtracted off to form $\Delta\varepsilon$, which is the energy gain with respect to the synchronous particle.

The phase change after an acceleration is $d\psi$ in general, and the synchronous part is also subtracted off in order to track the phase ϕ with respect to the synchronous phase. The identity

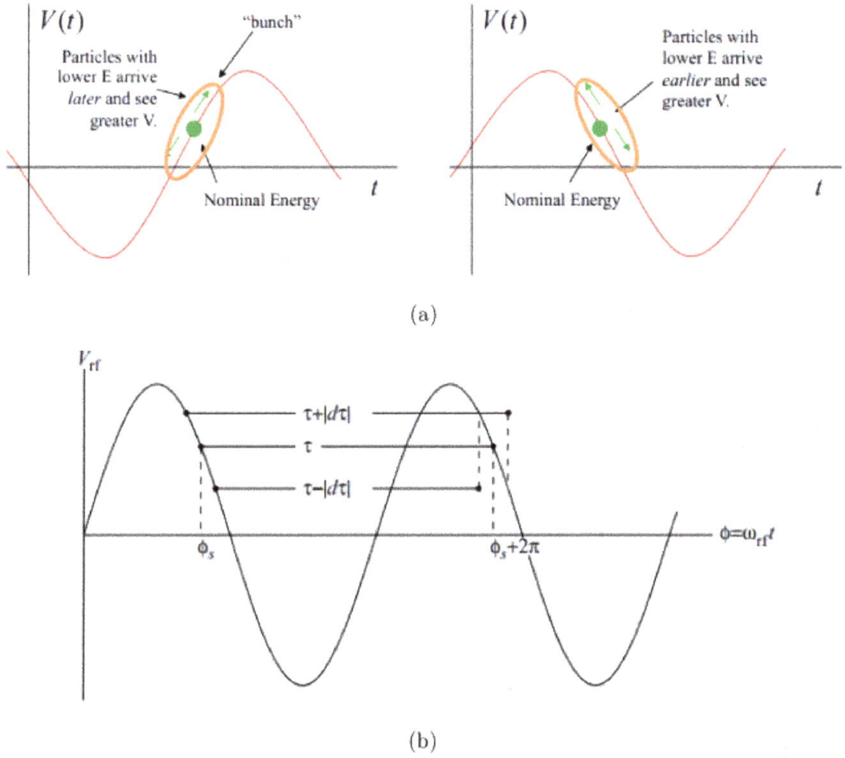

Figure 13.11: (a) Schematic of the region of phase stability both below and above transition energy. (b) Schematic showing — above transition — more energetic particles arriving earlier and sustaining less acceleration and vice versa, leading to phase stability. The phase of the particles changes with the next acceleration cycle.

$dp/p = (1/\beta^2)(d\varepsilon/\varepsilon)$ is also easily proven symbolically, similarly to what was illustrated in Figure 13.10.

$$\begin{aligned} d\psi &= \omega_o(\tau + d\tau) \\ \phi &= \psi - \omega_o\tau \\ d\phi &= \omega_o\tau(d\tau/\tau) = [\omega_o\tau\eta(dp/p)] = [\omega_o\tau\eta d\varepsilon]/(\beta^2\varepsilon) \\ \Delta\varepsilon &= \varepsilon - \varepsilon_s = qeV_o(\sin\phi - \sin\phi_s). \end{aligned} \tag{13.4}$$

The variable $\Delta\varepsilon$ is the energy gain of the asynchronous particle with respect to the synchronous one. This analysis of a single gap can be used because the RF, ω_o, and the synchrotron oscillation frequency are so different in magnitude that the synchrotron frequency can be regarded as constant.

For example, at the LHC the oscillations about the synchronous phase, with angular frequency, $\omega_{\rm syn}$ is 62 Hz, compared to the revolution frequency of 11 kHz and the RF of 392 MHz. For the Tevatron at 150 GeV, the ratio of synchrotron oscillation frequency to betatron oscillation frequency is about 0.002.

The variables are phase angle with respect to the synchronous phase, ϕ, and energy gain with respect to the synchronous particle's energy gain, $\Delta\varepsilon$. The oscillations mix particles in phase and the particles mix longitudinally in space as is visible in Figure 13.11. These changes are very slow in terms of the revolution frequency. For example, for the Main Ring, gaining 2 MV per turn, there are 47,899 turns per second of a gain of 95 GeV per second. A ramp to 400 GeV then would take about 4 sec.

To describe the motion, time, t, or number of turns, n, can be used as the independent variable with the relationship $dn = dt/\tau$. The previous analysis explored a single turn, which can now be considered as a continuous variable given the slow speed of the synchrotron oscillations.

$$\begin{aligned} d\phi/dn &= (\eta\omega_o\tau/\varepsilon_s\beta^2)\Delta\varepsilon \\ d(\Delta\varepsilon)/dn &= qeV_o(\sin\phi - \sin\phi_s) \\ d^2\phi/d^2n &= (\eta\omega_o\tau/\varepsilon_s\beta^2)(qeV_o)(\sin\phi - \sin\phi_s). \end{aligned} \tag{13.5}$$

For zero synchronous phase angle, the form of the equation is the same as the previously discussed pendulum with large angles, without the driving term. For small phase excursions, the synchronous phase can be subtracted, $\Delta\phi = \phi - \phi_s$, and the second order equation, using trig identities found in Appendix D, exhibits simple harmonic motion:

$$\begin{aligned} d^2\Delta\phi/d^2n &= \omega_{\rm syn}^2\Delta\phi \\ \omega_{\rm syn}(n) &= \sqrt{-\eta\omega_o\tau(qeV_o\cos\phi_s)/\varepsilon_s\beta^2}. \end{aligned} \tag{13.6}$$

For $\gamma > \gamma_t$, η is >0 and the synchronous phase must be within the limits: $\pi/2 < \phi_s < \pi$. Below transition the limits are $0 < \phi_s < \pi/2$ (Figure 13.11). A situation well below transition will be assumed for the numerical estimates which will be made but using a value for η which is applicable far above transition energy. The frequency shown in Equation (13.6) is the synchrotron frequency in units of turns, not time.

In order to adopt Fermilab conventions, the variables $y = \Delta\varepsilon/\omega_o$ in units eV*sec or $y = (Rcp/hc)(dp/p)$ will be used, with cp in eV. Time will be used as the independent variable, so that the differential equations transform, $dn = dt/\tau$, to

$$\begin{aligned} &d(\Delta\phi)/dt = Ay, \quad dy/dt = B[\sin(\Delta\phi) - \sin(\phi_s)] \\ &A = (hc/R)^2(\eta/\varepsilon_s), \quad B = qeV_o/2\pi h. \end{aligned} \tag{13.7}$$

For small phase differences and zero synchronous phase, the circular frequency is $\omega_{\text{syn}}(t) = \sqrt{AB}$ in inverse time units. For the Main Ring at 150 GeV, $\eta \sim (1/18.7)^2$, $\beta \sim 1$, $\phi_s = 0$ (no net acceleration), $\omega_o = 333$ MHz, and $\omega = 0.3$ MHz. Using the turns formulation, the circular frequency is 0.16 "Hz" per turn, $\omega_{\text{syn}}(t) = \omega_{\text{syn}}(n)/\tau$. Using the time variable, it is 770 Hz, much slower than the revolution frequency. The constants are $A = 2125/(\text{eV*sec}^2)$ and $B = 286$ eV. The region of stable phase space is called the RF "bucket". For the entire RF bucket, phase, width 2π for zero synchronous phase angle, the maximum value of y is $y_{\text{max}} = \pi\sqrt{B/A} = 0.72$ eV*sec (Figure 13.13). Changing variables, to dp/p, the maximum fractional momentum which is stably captured with no net acceleration is 0.26%. The bucket boundary is called the separatrix. There are "bunches" of particles inside the bucket. They may not populate the full allowable stable phase space.

The script "Phase_Stability" creates a plot of η and ω_{syn} as a function of γ, which is shown in Figure 13.12. Note that η is almost constant at high values of γ and the synchrotron frequency vanishes at transition. The value of η at high energies is used for the numerical estimates shown here but the synchronous phase is taken to be in the first quadrant for convenience.

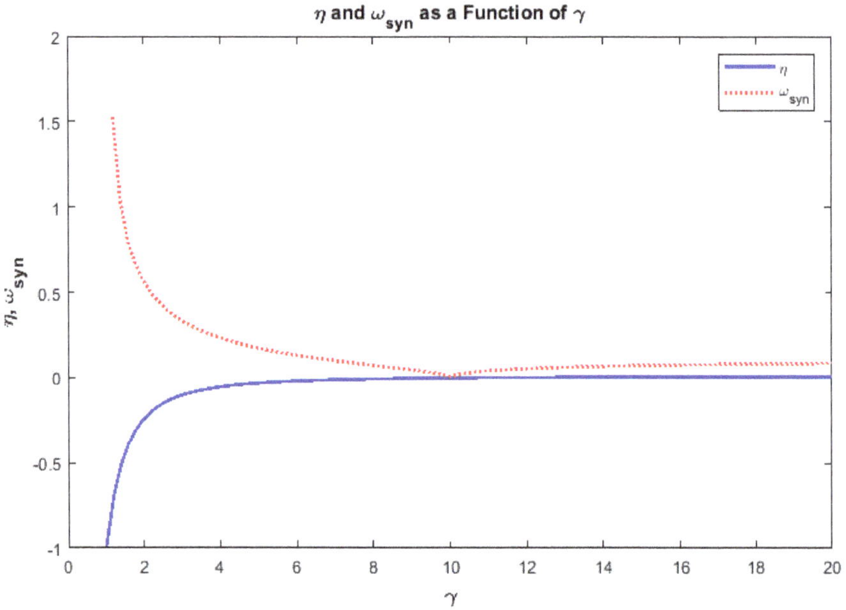

Figure 13.12: The behavior of η as a function of γ, where a transition γ is defined to be equal to 10. The synchrotron frequency initially falls as $1/\beta$, crosses zero at transition, and then grows slowly as η grows [Eq. (13.6)].

If V_o and ε_s are constant, a constant of the motion, analogous to total energy, can be defined. Since the energy gain per revolution is small with respect to the total energy, this approximation is useful.

$$
\begin{aligned}
&1/2(d\phi/dn)^2 + [\eta\omega_o\tau(qeV_o)/\varepsilon_s\beta^2](\cos\phi - \sin\phi_s) \\
&d\phi/dn = (\eta\omega_o\tau/\varepsilon_s\beta^2)\Delta\varepsilon, \quad \text{Eq. (13.5)} \qquad (13.8) \\
&\Delta\varepsilon^2 + (2qeV_o\varepsilon_s\beta^2)/\eta\omega_o\tau](\cos\phi - \sin\phi_s).
\end{aligned}
$$

The script "Phase_Stability", employing user-supplied values for the synchronous phase angle and the y value, calculates the maximum y and phase that the bucket can contain and the bunch "area", which is simply taken to be the rectangular area between the maximum limits.

A plot of the evolution of the phase and energy gain as a function of time for the parameters supplied in Figure 13.13 is

```
>> Phase_Stability
  look at phase stability for r.f in the MR ar 150 GeV

Enter Synchronous Phase (degrees): 0
Enter (phio = 0), yo ~ (0.1 - 0.6) eV*sec: 0.73
For phis=0, max y Bucket (eV*sec) = 0.733725, dphi Bucket = 2*pi
Bunch - maxy (eV*sec) = 0.73106, max and min phi (rad) = 2.94703 , -2.9358
Approximate Bunch Area = 8.59154 (eV*sec*rad)
```

Figure 13.13: Printout from the script "Phase_Stability" for user-supplied values of the initial y and the synchronous phase angle.

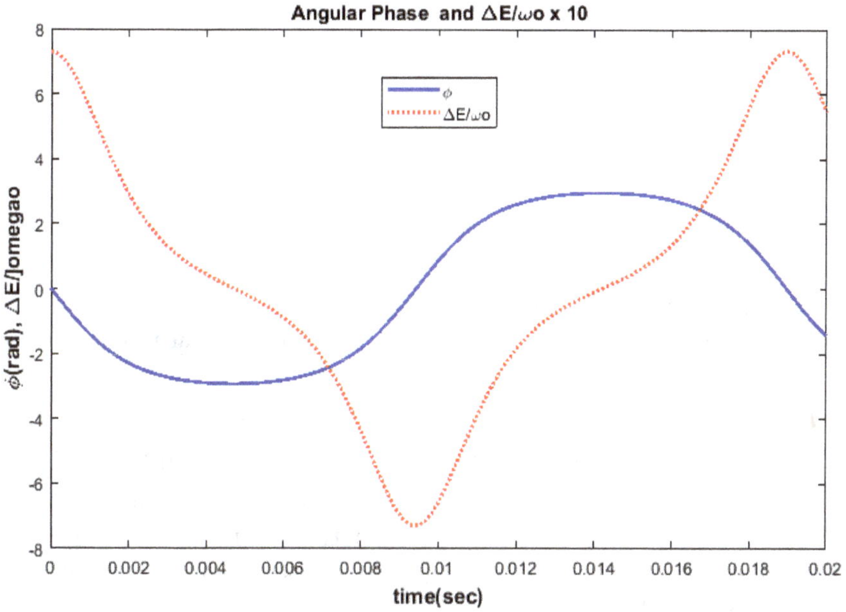

Figure 13.14: Behavior of phase and $\Delta\varepsilon$ near the region of maximum stable energy gain as a function of time for $\phi_s = 0$ and both $y_o = 0.5$ and $y_o = 0.73$.

shown in Figure 13.14. The behavior is far from sinusoidal because the parameters are near the stability limits as calculated in the printout. The utility "ode45" is used to solve the nonlinear equations of motion. A "movie" of 100 frames is also displayed using this script to explore the time evolution of the system. The movie illustrates that as y_o increases, the time spent near the maximum and minimum phase increases.

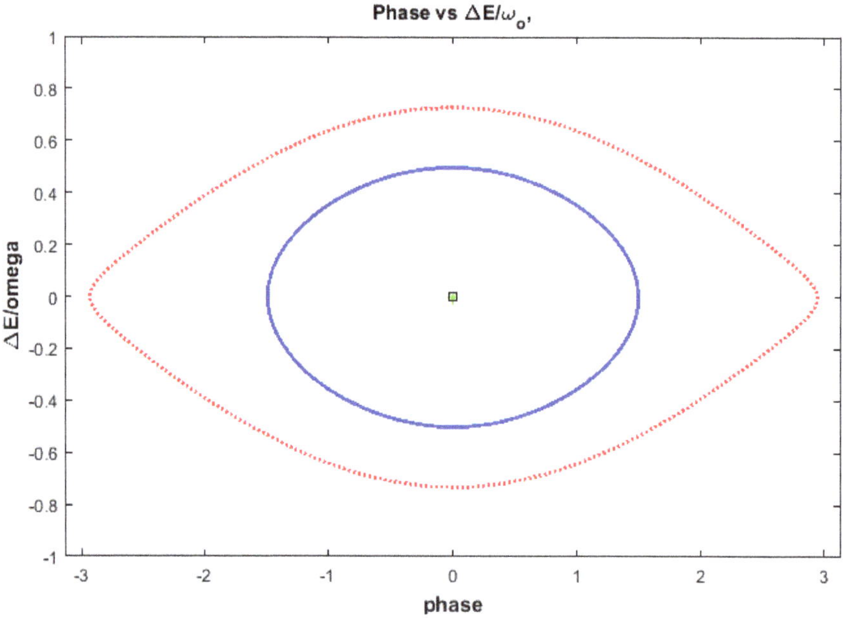

Figure 13.15: Behavior of phase vs energy gain for a synchronous phase of zero and $y_o = 0.5$ (solid) and $y_o = 0.73$ (dotted). The latter is near to the limit for stability of oscillations.

The phase and energy gain are shown as contours in Figure 13.15 as an alternative to the time development shown in Figure 13.14. When the synchronous phase is zero, or there is no net acceleration, the stable phase covers a 2π range when y is the maximum allowed stable value. The circular shape for smaller values of y indicates that the small y version of the equations of motion yields simple harmonic motion, as was the case for the classical pendulum. The phase excursion increases as y increases. At some point a sharp transition to unstable motion occurs.

When the synchronous phase is nonzero, indicating net acceleration, the shape of the phase vs y changes, it is still centered on the synchronous phase but the stable values of both the phase and y shrink, indicating lessened phase space acceptance and a smaller dp/p. In Figure 13.16 the synchronous phase is chosen to be 30°, and the resulting, maximum y value is reduced to about 0.38 from

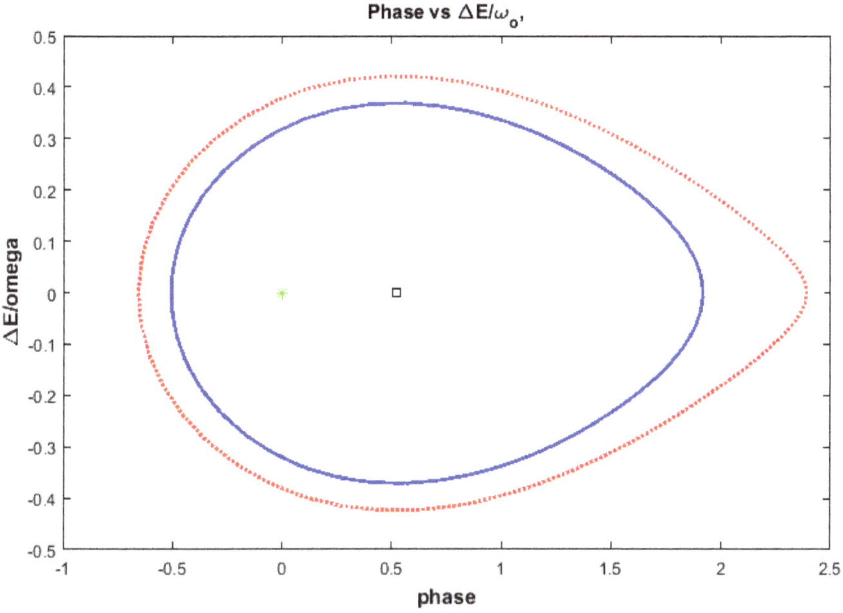

Figure 13.16: Behavior of the synchrotron phase and y near the region of maximum stable energy gain as a function of time for $\phi_s = 30^\circ$ and $y_o = 0.32$ (solid blue) and 0.38 (dotted red). The origin appears as a green *, while the synchronous orbit is shown as a black square.

0.73 at 0°. The bunch area also has a reduced phase range, with a maximum of about 2.57 radians as opposed to the 2π value observed previously. As the beam accelerates, the bunches narrow in dp/p and in the stable phase range.

There are unstable regions of the phase space also, just as there are unbounded Keplerian orbits. The border between stable and unstable behavior is very sharp. In Figure 13.17 the plot of phase vs y is for zero synchronous phase angle, and with $y = 0.73$ and 0.76. The latter "orbit" is unbounded, rather like the elliptical and hyperbolic orbits in Keplerian orbital motion. The user can explore both bounded and unbounded motion using "Phase_Stability". The unstable particles will fall out of the "buckets" and be lost to the accelerated beam.

A MATLAB "app" was written, — "Accel_RF", which allows the user to select the time interval to display, the synchronous phase,

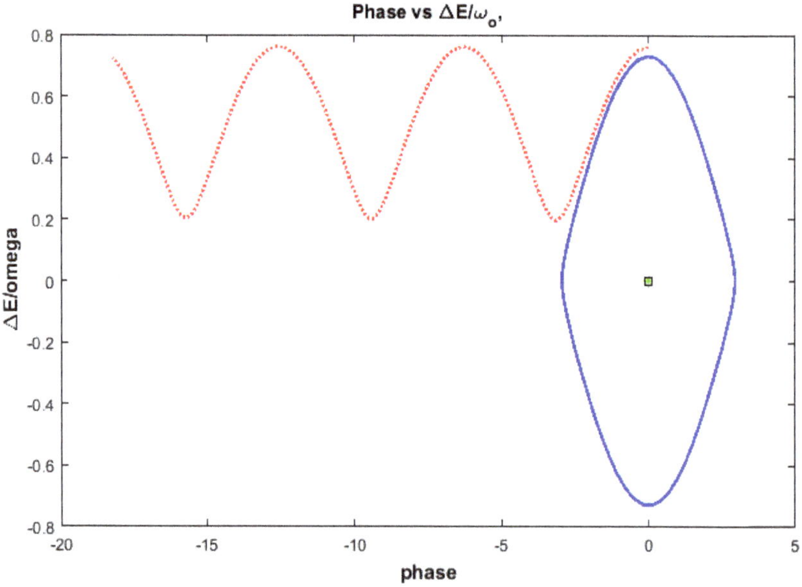

Figure 13.17: Phase vs y orbits for zero synchronous phase and $y = 0.73$ (solid blue) and 0.78 (dotted red), which is unbounded.

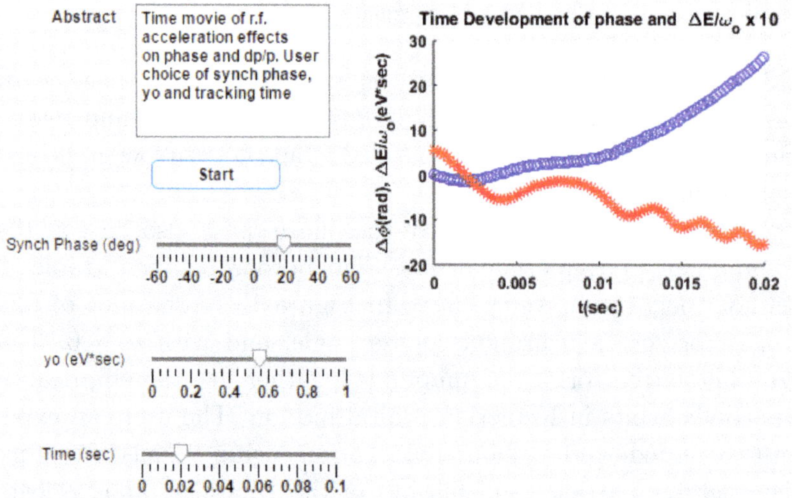

Figure 13.18: Display of the last frame of a movie generated by "Accel_RF". The specific case has unbounded behavior.

and the initial value of y_o, which is related to the dp/p acceptance of the acceleration cycle. A "movie" of the development in time is made. A specific plot of unbounded behavior appears in Figure 13.18. The user is encouraged to "play" with the variables and to explore both stable and unstable behaviors. It seems clear that a planned change in the RF could be used to extract the beam in a controlled fashion.

The unstable behavior has a sharp boundary. The result of searching for that boundary using the script "Phase_Stability_2" appears in Figure 13.19. The script searches over a grid of synchronous phase and y_o in order to find the unstable regime and then

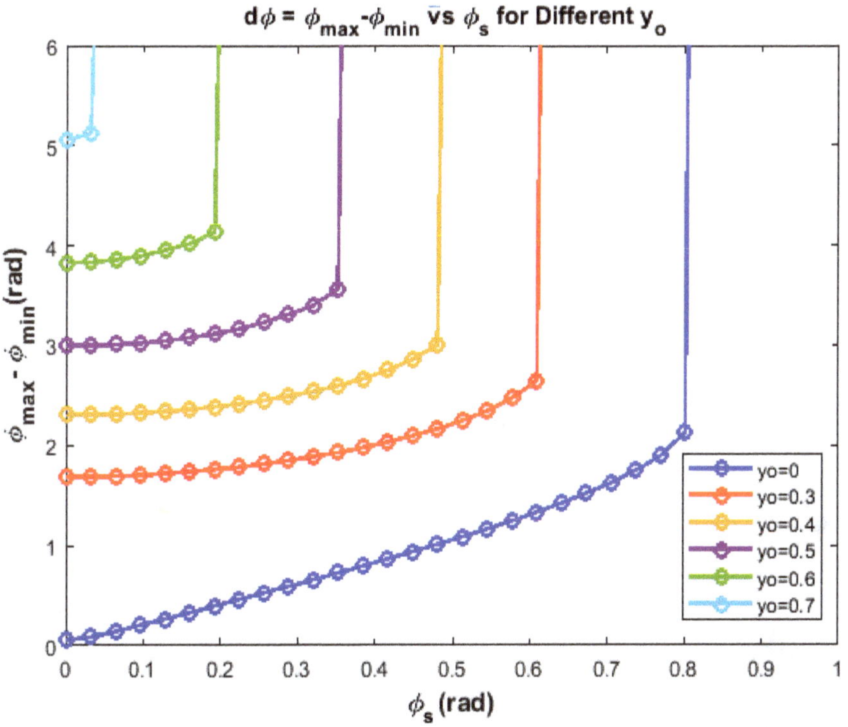

Figure 13.19: Dependence of the extent of stable phase oscillations as a function of the synchronous phase for different values of y_o. Unstable behavior is indicated by the vertical lines in phase acceptance for different y_o, which arise as the magnitude of the synchronous phase angle increases.

calculates the region in phase where the motion is stable. The stable region of synchronous phase shrinks with increasing y_o, or dp/p. These points can be explored individually using the app "Accel_RF", which can be a useful exercise.

For insertion and extraction, one wants as small a beam as possible. The bunches should also be synchronized, so that successive accelerator stages are normally RF-phase-locked. In general, when matching beams, $V_o \cos \phi_s / h\eta$ should match. The Fermilab Booster RF at injection has a frequency of 37 MHz which rises to 53 MHz at 8 GeV energy. The Main Ring operates at the same RF, 53 MHz.

Chapter 14

Medical Applications and Light Sources

This new knowledge has all to do with honor and country but it has nothing to do directly with defending our country except to help make it worth defending.

— **Robert Wilson**

We must not forget that when radium was discovered no one knew that it would prove useful in hospitals. The work was one of pure science. And this is a proof that scientific work must not be considered from the point of view of the direct usefulness of it. It must be done for itself, for the beauty of science, and then there is always the chance that a scientific discovery may become like the radium a benefit for humanity.

— **Marie Curie**

As shown in the preface, radiotherapy is the largest use of accelerators, followed by ion implantation employed in the semiconductor industry. The use by high energy physics is minuscule by comparison. Nevertheless, there are medical applications of accelerators whose primary goal is basic research. Two examples from Fermilab are explored in this chapter.

The Fermilab Neutron Therapy Facility (NTF) was first used in 1975. A schematic of the neutron beam used for therapy appears in Figure 14.1. Protons from the Linac are extracted with a pulsed dipole and transported to a target using quadrupoles and dipoles. At the end of that short beamline, neutrons are produced and then collimated using shaped absorbers with shapes personalized to the needs of the particular patient.

Neutrons, being uncharged, and having a fairly long mean free path, are preferentially used in the treatment of deep-seated tumors. When neutrons interact they create a variety of ionizing, short range

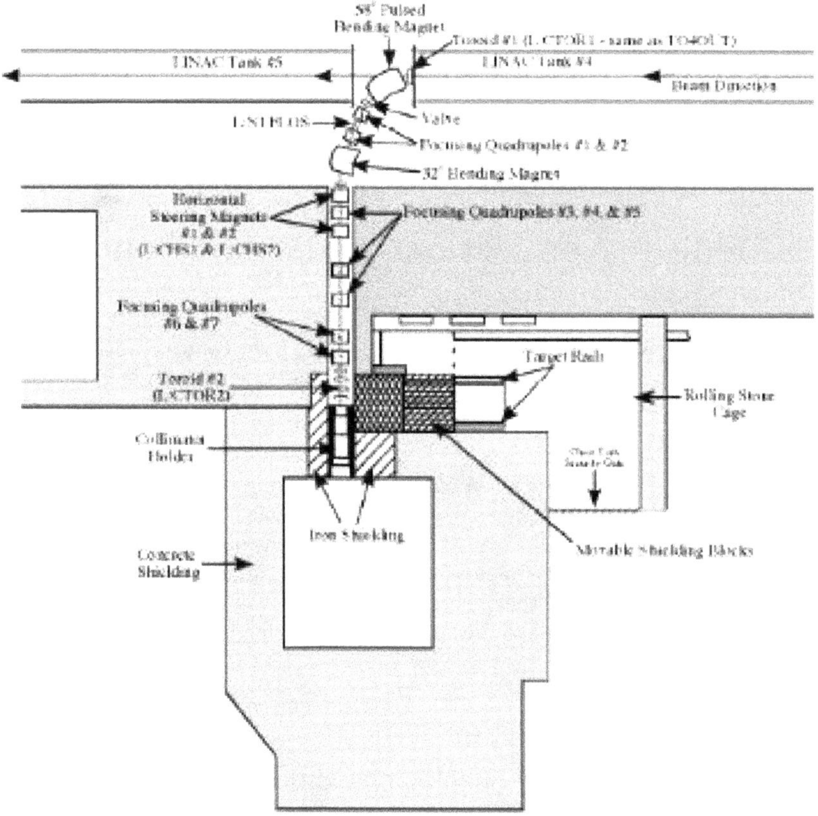

Figure 14.1: Use of the Fermilab Linac to make a therapeutic neutron beam employing collimation.

particles which interact with the tumor and deposit significant energy in it.

The use of neutrons in cancer therapy is limited to specific types of disease. The Fermilab NTF has been used for many years in the treatment of these specific cases.

Accelerators have many applications, and medical applications are among the most important. Their use relies on some interesting physics. First, the script "Min_Ioniz" is used to explore the electric field of a fast-moving particle. The dialogue for this script appears in Figure 14.2.

```
Electric Field Contour, Point Charge
z = instantaneous position to observation pointCharged particle Moving Along z Axis
 Velocity w.r.t. c is b, Observation point at (b,0,0)
 Unit for E is Exmax = e/b^2
Span for t is +- 5 units of b/v
 Enter Velocity of Charge beta: 0.9
Enter Velocity of Charge beta: 0.95
```

Figure 14.2: Printout of the script. The nonrelativistic field at the point of closest approach is just e/b^2.

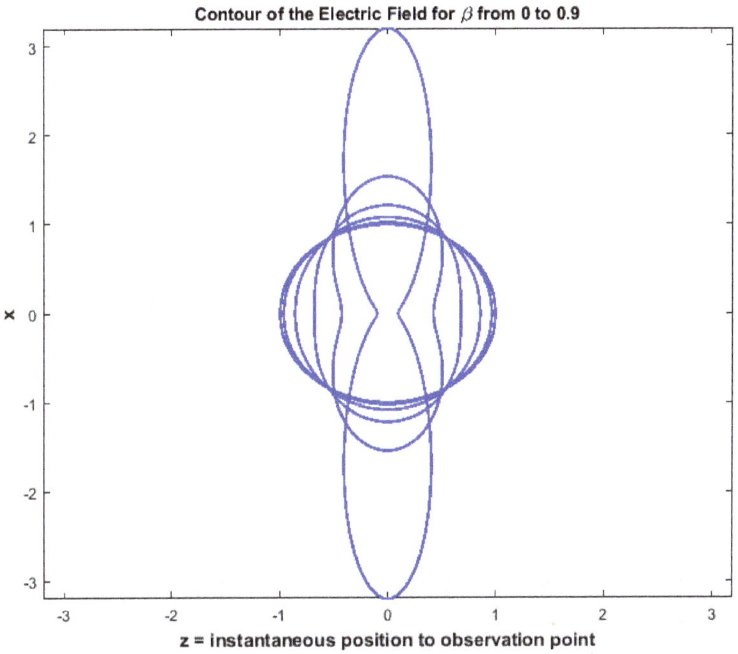

Figure 14.3: The electric field of a point particle. The maximum field at the point of closest approach exceeds the classical value.

The output of the script gives contours of the electric field for different $\beta = v/c$ values and plots of E_x and E_z as a function of time. Only E_x (transverse) has a nonzero time integral by symmetry. The electric field is also shown as a movie as it develops in time for a user-chosen value of the velocity. An example appears in Figure 14.4. The user can explore how the field develops as the particle velocity approaches c.

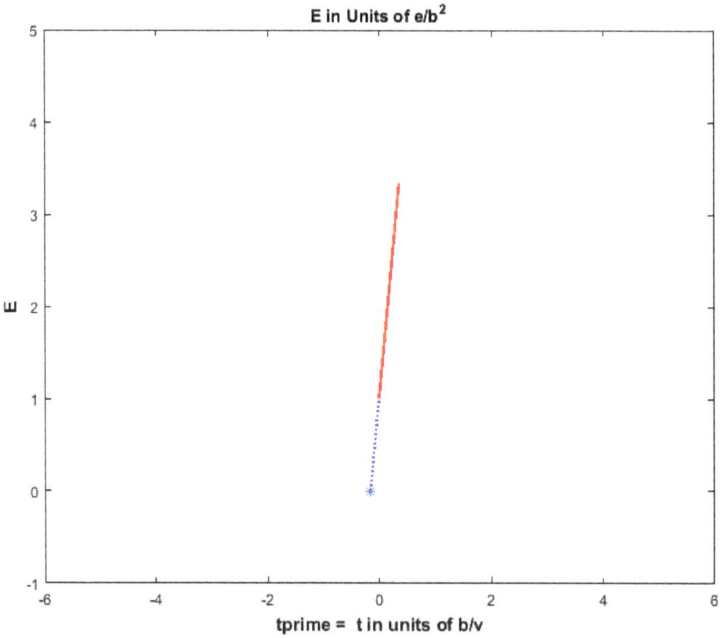

Figure 14.4: Frame of a movie showing the electric field vector (red solid line) at the observation point as a function of time. The location of the line from the charge to the observation point is shown as the blue dotted line.

The electric field shrinks in time and grows in x such that the momentum impulse approaches a constant. Since that impulse transfers energy and causes ionization of the medium, relativistic particles give constant ionization, called the "minimum ionizing particle", or "mip". This fact has far-reaching therapeutic implications. The user can play the movie for any value of β in order to develop a feeling for the electric field that results.

Because of the relativistic effect, it is expected that high energy charged particles lose a small and constant amount of energy in traversing material. However, at lower velocities, the particle spends more time close to the point of closest approach, $b : t \sim b/v$. Since the field at that point is strongest due to the inverse square distance behavior of electromagnetism, ionization should be heaviest at that point. This behavior is explored in the script "Energy_Loss", as illustrated in Figure 14.5.

```
>> Energy_Loss
  Range and KE of a slow moving proton - medical accelerator

|

Enter the Inital Kinetic Energy of the Proton (MeV): 100
Range in Water in cm = 5.35019
```

Figure 14.5: Dialogue for "Energy_Loss", where the user selects an incident energy.

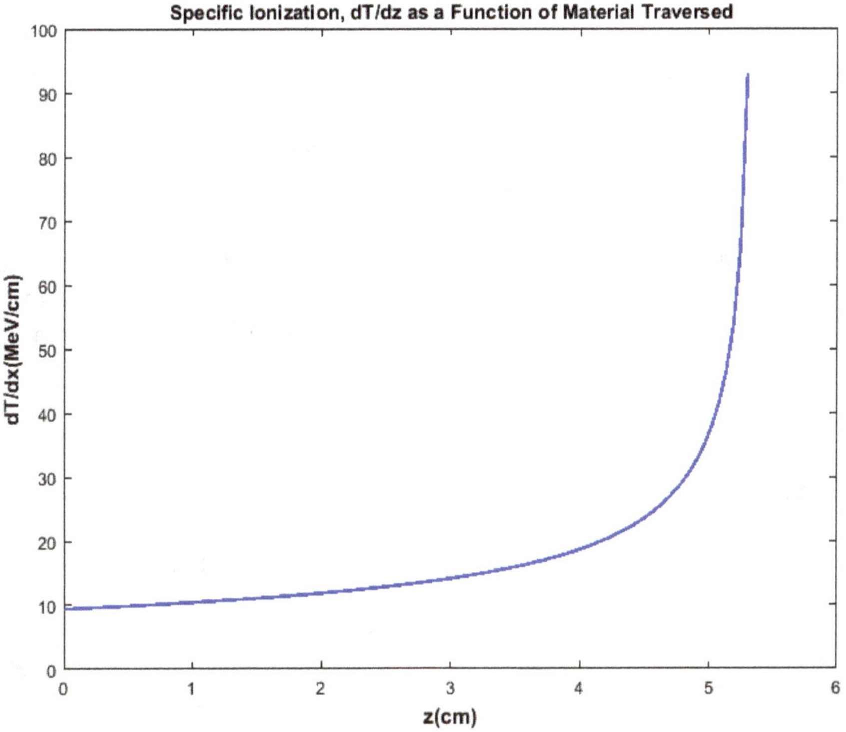

Figure 14.6: Energy loss as a function of depth for a 100 MeV proton incident in water.

Patients being largely water, the script uses water as the ionized medium. A proton loses energy, slowly at first, but faster as it slows down. It stops, having lost all its energy at a distance called the "range". A plot of the proton kinetic energy as a function of water depth is shown in Figure 14.6. The peak in deposited energy per unit length at the end of the range is called the "Bragg peak" and it has clear medical applications.

A comment here on notation. The standard formulation of energy loss is dE/dx. However, the energy E is actually the kinetic energy, $T = \varepsilon - m$, and dx refers to motion along the direction of motion, or z. Further, the quantity is normally quoted in MeV per gm/cm^2 (cgs units are the standard, here not MKS), so a better notation might be $dE/dx \rightarrow dT/d(\rho z)$.

The proton also multiple-scatters in the material, which smears out the sharp peak shown in Figure 14.6. A measured energy deposition of a 205 MeV proton in a material with density 0.97 gm/cm^3 (near water) appears in Figure 14.7. The range is 26 cm and the maximum energy deposit is about four times the minimum.

This energy loss behavior is of great use in medicine, since a dose of radiation can be targeted by choosing the extracted beam energy so that the deposited energy is preferentially large at a tumor site, thus sparing the intervening tissue in the case of deeply sited tumors. Fermilab has been involved with the Loma Linda Medical Center and

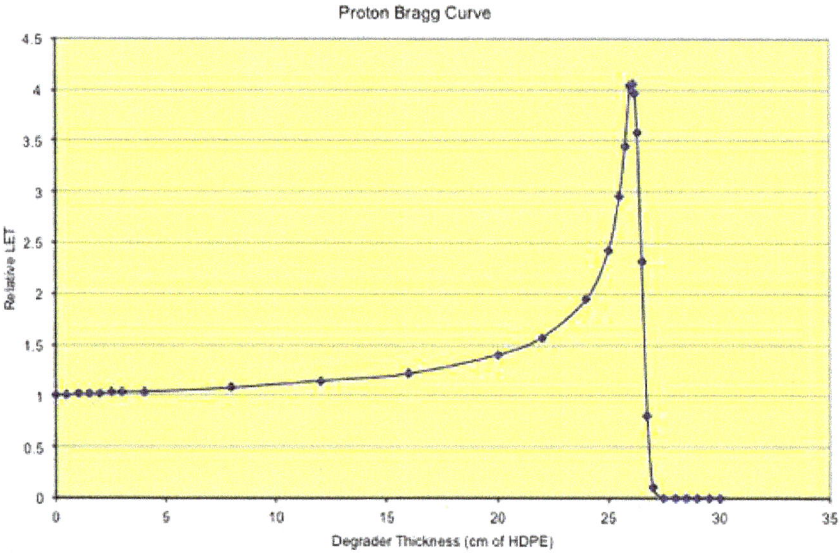

Figure 14.7: Energy deposit in high density polyethylene as a function of depth in the material.

Figure 14.8: Photo of the Loma Linda accelerator.

they have jointly designed and built an accelerator of variable energy for patient treatment. A photo of the machine appears in Figure 14.8. Loma Linda began treating patients in 1990 — the first such facility to operate in the US.

There are now many such facilities worldwide for treatment of patients with proton beams. Indeed, there are also several facilities using heavier ions as beams for therapy. To define a scale, a unit of radiation is the rad. Natural background sources are about 0.1 rad annually, while a therapeutic dose might be 10 rad per session or a deposition of 0.1 J/kg. Taking water as the density, that dose is about 6×10^8 MeV in a cc of water. A mip deposits about 1.5 MeV/(gm/cm^2) or 1.5 MeV/cm in water. This means that a dose of about 400 million high energy protons would be needed. These numbers are only order-of-magnitude but they convey a rough idea of the doses involved.

The second-commomest use of accelerators is in the semiconductor industry. Electrostatic devices with voltages of order 100 keV are employed, but there is a wide range of energies in use. They accelerate ions to create dopants at the required small densities, about 10^{16} ions/cm^2. The implantation depth is set by the range of the ions in the substrate silicon and is typically small, of order 10 nm, which is accomplished rapidly with ion beams with currents of order 100 μA.

As seen in the preface, another important use of accelerators is in the generation of intense sources of photons. The radiation of accelerated electrons was first seen in circular electron accelerators and is called synchroton radiation in consequence. Indeed, the energy loss for the electrons grows very rapidly with energy and becomes a limiting factor in circular electron accelerators. The Large Electron–Positron collider (LEP) at CERN was perhaps the last of the high energy circular accelerators because of the synchroton power loss problem.

Linear electron–positron colliders do not have a severe radiation loss problem and are the current candidates for the next-generation electron–positron collider, such as the International Linear Collider (ILC). In hadron colliders the energy loss by radiation is reduced by an approximate factor of $(2000)^4$ for the same-energy circular collider. Nevertheless, the beam pipe heating due to proton radiation at the CERN Large Hadron Collider at very high luminosity also becomes an issue. At a high-enough energy all stable particles are relativistic and radiate a significant amount.

A map of some of the accelerator "light sources" is shown in Figure 14.9. There are many such facilities, often using electrons of several GeV to produce intense beams of photons at keV energies and below. These facilities are heavily subscribed by researchers in the life sciences, chemistry, materials science, and condensed matter physics.

In the case of nonrelativistic acceleration, the Larmor formula states that the radiation pattern is dipole and that the power depends on the square of the acceleration of the charge. In the case of relativistic radiation, there are changes to that expectation, ones that are covered in many textbooks on electromagnetic theory. Here, only the results will be quoted. A basic result is that the

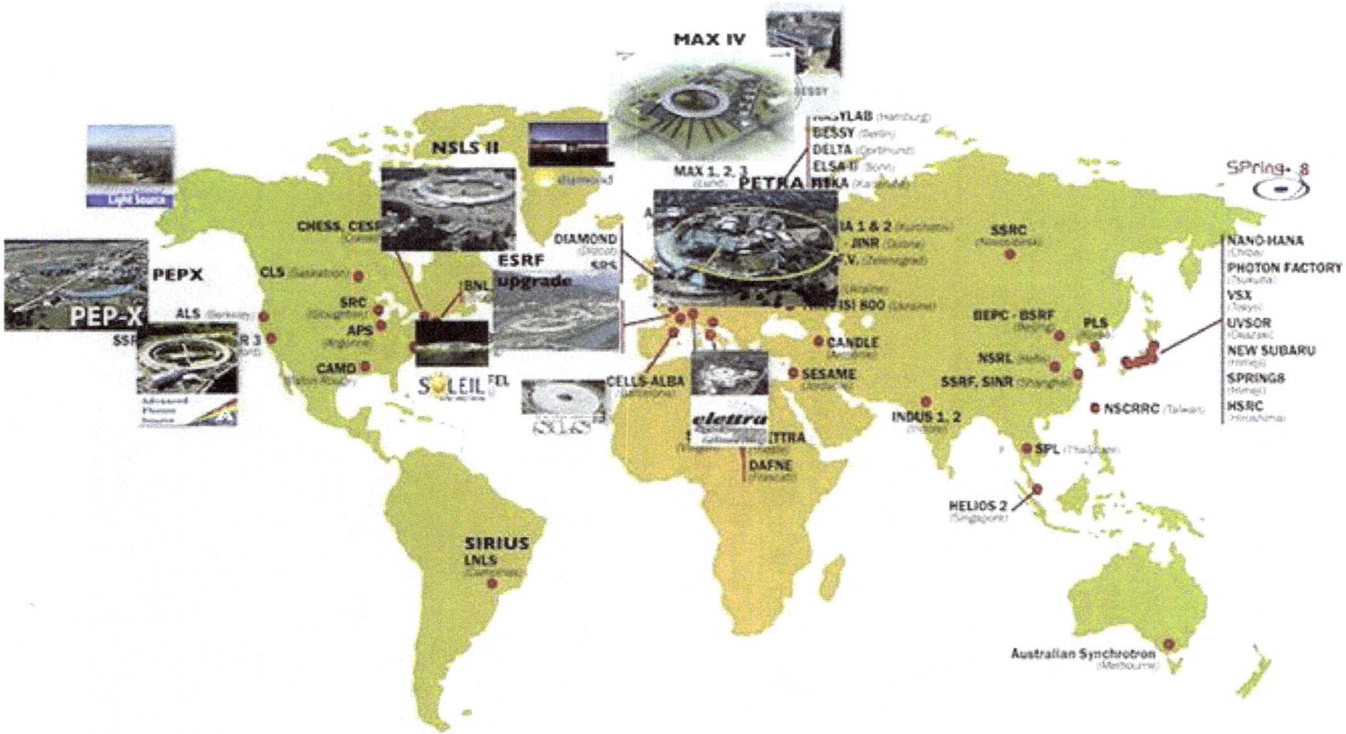

Figure 14.9: World map showing some of the many light sources operating at present. There are also several "free electron laser" facilities starting operation which provide very intense photon beams in the x-ray energy range.

change in momentum and the change in velocity are related as $dp = m(1/\gamma^3)d\beta$, which reflects the fact that β cannot exceed 1 while p can increase indefinitely. This result is easily found symbolically using the three MATLAB command lines: p = b/sqrt(1-b^2), dp = diff(p,b), and pretty(simplify(dp)).

For acceleration perpendicular to the direction of motion of a charged particle, the power, P, goes as $P \sim \gamma^4\dot{\beta}^2$(the dot defines the time derivative), which scales, for circular motion, as γ^4/ρ^2, since acceleration goes as $1/\rho$ for circular motion. These results hold in general and apply to any accelerated charge. A specific case where the emitted photons are used experimentally is the situation with light sources.

The acceleration is typically provided by an "undulator". A schematic of such a device appears in Figure 14.10. Rare earth permanent magnets are often used to supply the fields. The electron beam is assumed to be a pencil beam incident along the z axis. It encounters a magnetic field along the y direction which is modulated by the periodicity of an undulator wave vector, k_u. This field causes

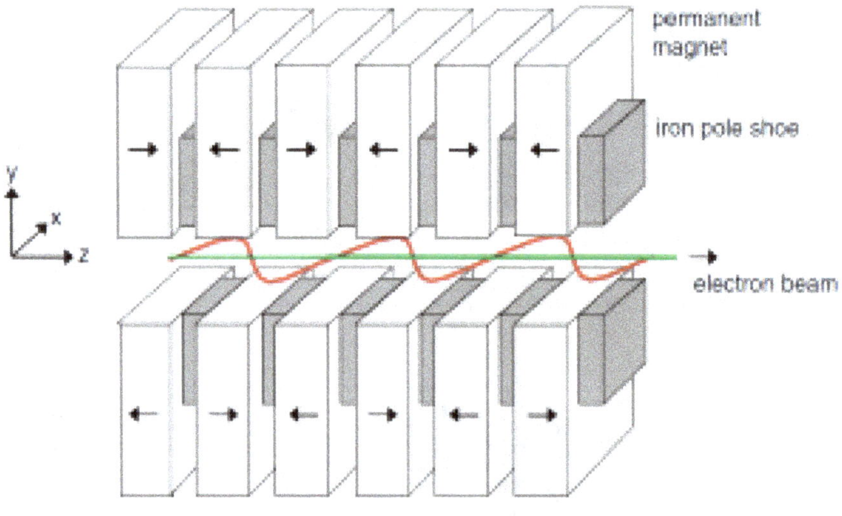

Figure 14.10: Schematic of an electron beam "undulator", with the accelerated motion indicated in the (x, z) plane. The magnetic field direction is shown as arrows on the permanent magnets.

an acceleration of the electron beam transverse to the beam direction, here taken to be the x direction.

The modulated magnetic field results in an acceleration with the periodicity of the undulator wave vector ($q = 1$ assumed here).

$$\begin{aligned} B_x &= B_o \sin(k_u z) \\ \vec{a} &= d\vec{v}/dt = -e(\vec{v}x\vec{B})/(\gamma m). \end{aligned} \tag{14.1}$$

Very approximately, the dipole radiation emitted in the beam rest frame is thrown forward into an emitted cone in the lab frame with typical angle $\theta = 1/\gamma$. This is the well-known "searchlight effect" of relativity. For a fixed laboratory observer the angular collimation means that there are short bursts of photons in time. That leads to an emission spectrum containing frequencies up to a maximum critical value of $\omega_c = 3c\gamma^3/(2\rho)$.

The energy of the emitted photons, $\varepsilon \sim k$, is also enhanced by relativistic factors over the rest frame value of k_u given by the accelerating field, $k \sim k_u(2\gamma^2)$. The angular distribution of an isotropic beam in a rest frame transformed into the lab frame is explored in the script "Rel_Angles". The Lorentz transformation for photons of energy and the z component of momentum is differentiated symbolically using "diff" to find the Jacobian transformation from the electron rest frame to the laboratory frame, as shown in Figure 14.11.

```
>> Rel_Angles
Lab Angular Distribution - Isotropic Beam Radiation

ans =

g^2/(b*cost - 1)^2

Searchlight Angle at beta = 0.9  24.9747 degrees
```

Figure 14.11: Printout of the script "Rel_Angles", showing the lab angular distribution.

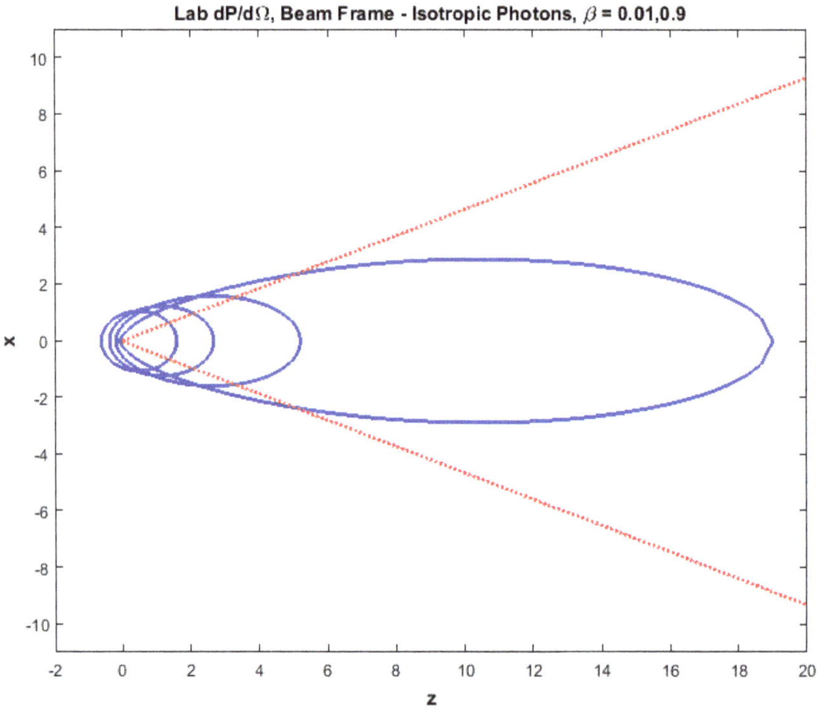

Figure 14.12: Contours in (x, z) of an isotropic rest frame distribution for different β values from 0.01 to 0.9 (solid) and the searchlight angle (dotted) for $\beta = 0.9$.

A plot of the angular distribution appears in Figure 14.12, where $dP/d\Omega \sim (1/\gamma^2)/(1 - \beta\cos\theta)^2$. Also plotted is the approximate "searchlight" angle, which indicates how the isotropic distribution seen at small velocity evolves into a strongly forward-peaked distribution. Since a 2 GeV electron beam has a γ factor of 3914, it is clear that the photons are very forward-peaked along the z axis. The effects of relativity mean that a wavelength of order cm causes a laboratory photon emission with wavelength of order nm due to the $1\gamma^2$ factor which arises from time dilation and length contraction, $\lambda_\gamma \sim \lambda_u/2\gamma^2$.

Numerically for a 2 GeV electron with a radius of curvature of 2 m, the critical energy, $\hbar\omega_c$, is 1.7 keV, where the Planck constant is $\hbar = 6.58 \times 10^{-16}$ eV*sec. The searchlight angle is 0.26 mrad.

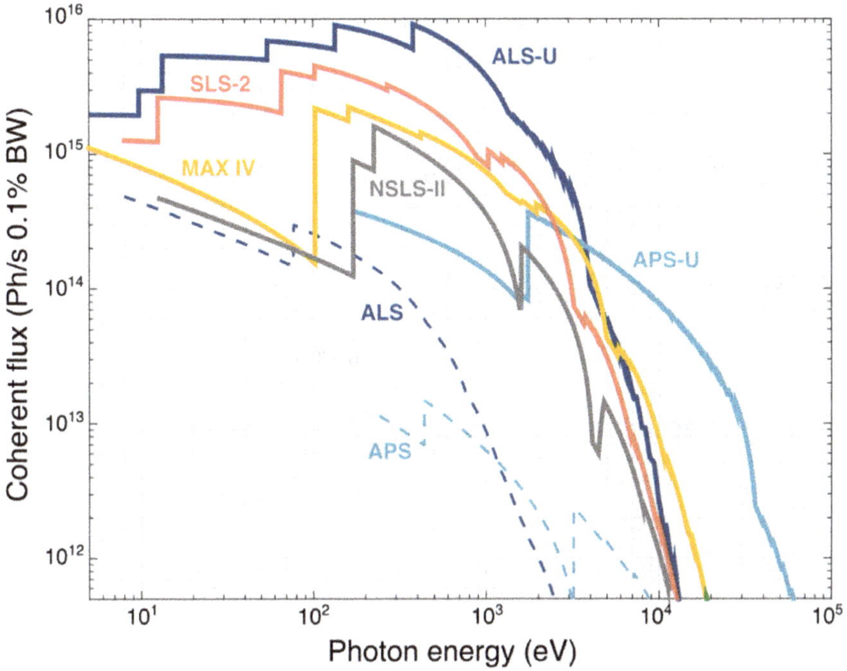

Figure 14.13: Energy flux of emitted photons from several light sources. The characteristic $1/\varepsilon$ falloff for high energy photons is evident. Typical photon energies are in the keV range.

The power spectrum of photons of energy ε_γ emitted incoherently by a current, I_e, of electrons of energy $m\gamma_e$ is very approximately

$$P \sim I_e B_o^2 \gamma_e^4 / \varepsilon_\gamma. \tag{14.2}$$

The detailed spectrum and power level is shown in Figure 14.13 for a variety of light sources. The falloff with photon energy is in approximate agreement with Eq. (14.2). Typical energies are in the keV range, as was already estimated.

The emission of synchrotron radiation, as the name implies, occurs during acceleration of relativistic particles, such as electrons, in circular accelerators. The emission of photons by the electron beam is a perturbation on the beam. Stable operation needs to be insured. Indeed, at extreme energies this radiative energy loss by the beam must be supplied by the external RF system and imposes a fundamental limit on circular accelerators. The power radiated in GeV/sec for an electron of energy ε is

$$P = c_{sr}\varepsilon^4 c/2\pi\rho^2. \tag{14.3}$$

The dependence on energy is quartic, as expected from the power considerations mentioned above. The numerical factor for an electron synchrotron is $c_{sr} = 8.84 \times 10^{-5}\,\text{m/GeV}^3$. The total power depends on the beam current. Numerically, for the LEP collider at CERN, with a beam current of 6 mA and a beam energy of 100 GeV the RF must supply 18 MW. Clearly, higher energies become economically problematic unless the radius of curvature is increased, which would require a new and expensive accelerator ring.

Chapter 15

Summary

> Before I came here I was confused about this subject. Having listened to your lecture I am still confused. But on a higher level.
>
> — **Enrico Fermi**

This text has made a simple survey of topics in beam and accelerator physics. Only the first order, linear, effects have been covered in any detail. Connections to classical mechanics have been given since the mathematical tools are the same and many accelerator objects and systems have analogues in classical mechanics.

The aim has been to be broad but not terribly deep. Many topics are covered using a single coherent tool — MATLAB. The student will, however, be exposed to many topics in a visual and hands-on fashion. Typically, the MATLAB scripts have "movies" which map out the dynamical evolution of a particular problem. Student input to vary the parameters of a problem comes through command line inputs using the "Menu" facility or by using "sliders" and other tools in an "app" script.

The scripts are integral to the text and are there to be run and explored. The scripts themselves are available to be read and modified. To that end, the MATLAB "live" option for the script can be used to track the script execution line by line. Many "comment" statements are made in the scripts in order to explain what is being evaluated at each line of code. A few "live" examples are supplied for the student to experiment with. It is hoped that the reader can go forward and write entirely new MATLAB code — code which is more simple and elegant than that made available here.

In this fashion the student is exposed to examples in both beam and accelerator physics. The main focus is on hadron accelerators, because of the background of the author. However, some examples in electron light sources and electron–positron colliders are given. Figures and formulations have been shamelessly imported from those appearing in lectures given at the US Particle Accelerator School. Indeed, the author lectured at that school in the past. The goal has been to introduce the concepts and principles of beam and accelerator physics in a particular visual and intuitive fashion. It is hoped that this goal has been, at least partially, met.

Appendix A

Inputs for "Plot"

A collection from the command line "help plot" showing the colors, the symbols, and the line types associated with the MATLAB utility "plot". For the scripts used in this text, the utility "plot" is the most-used display. Indeed, for the "app" scripts, "plot" is the only supported graphic. This simple two-dimensional display is, however, often sufficient for "app" scripts. Full documentation can be obtained using the search window in the Command Window taskbar.

b	blue	.	point	-	solid
g	green	o	circle	:	dotted
r	red	x	x-mark	-.	dashdot
c	cyan	+	plus	--	dashed
m	magenta	*	star	(none)	no line
y	yellow	s	square		
k	black	d	diamond		
w	white	v	triangle (down)		
		^	triangle (up)		
		<	triangle (left)		
		>	triangle (right)		
		p	pentagram		
		h	hexagram		

Figure A.1: Result of a "help plot" command line query. The columns show the possible colors, point symbols, and line formats.

Appendix B

Live and App Designer Scripts

The basic MATLAB script is a ".m" file. It typically uses the utility "input" to request parameter input from the user via the command line, prints to the command window, and makes plots to the figure window. Choices to be made often use the "Menu" utility to start a dialogue, end it, or request another type of input. This procedure is a very command line, old school, way to execute a program. More recently, MATLAB has created other options. The GUI option is very useful, but a bit cumbersome for this text because it uses two distinct files: a script itself and a figure file. Another option is called the "live script". Using the general help and search, much documentation is available (in particular a video).

In the Editor the "new" tab has the option to make a live script. It creates a file with the ".mlx" extension. This extension allows for additional commands beyond the standard .m file suite. For example, "Insert" allows the script writer to add both text (with formatting) and images. One can also add equations as text/image. The code itself is the same as for standard ".m" files, and the "Live Editor" is almost the same as the "Editor". The script can be run using the Run button and break points can be inserted as wished in order to enhance debugging capabilities.

A first example from the text is the live file for the motion in a uniform electric field, "Uniform_E_Motion.mlx".

One great advantage of this type of file is to split the Editor screen into code and output. The cursor can then be run down the code to execute the commands sequentially and examine the output on the right screen. This option will be used only occasionally in scripts describing the beams and accelerators in this text. It is very

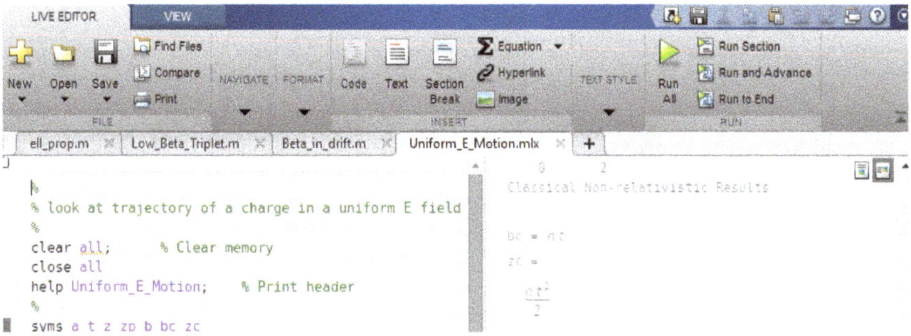

Figure B.1: The first "live script" to appear in the text with a split screen display. The script proper appears on the left side, while the output appears on the right.

easy to take a standard .m script and convert it to a .mlx script, which makes the choice of script type a matter of taste.

There is also an option in MATLAB to make "apps". These are a stripped-down version of the more elaborate MATLAB GUIs, but all in one file, script plus figure(s), rather than the two distinct files required in the GUI. However, the graphics is limited to simple plots. Often this is not a major limitation.

The documentation is extensive for apps in MATLAB, and many examples exist. A video is available. The command window is used to first invoke "appdesigner".

There are four distinct app windows which are created; the component library, the design view/code view toggle, and the component browser and the component properties. Text input can be made for a longer explanation of the script, using "Textedit".

The created ".mlapp" file is run using the Run button. Components are dragged from the component library to the design view window and customized there. Callbacks define in the code view what operations are made when a component (right click needed) changes, for example. Data is plotted using "UIAxes", and numerical input comes from "sliders" or other input objects. Numerical output can drive "gauges" to examine the results of the calculations. The plot uses plot(app.UIAxes,app.xdata,app.ydata). Other plot commands, such as "title", are the same as in ".m" files.

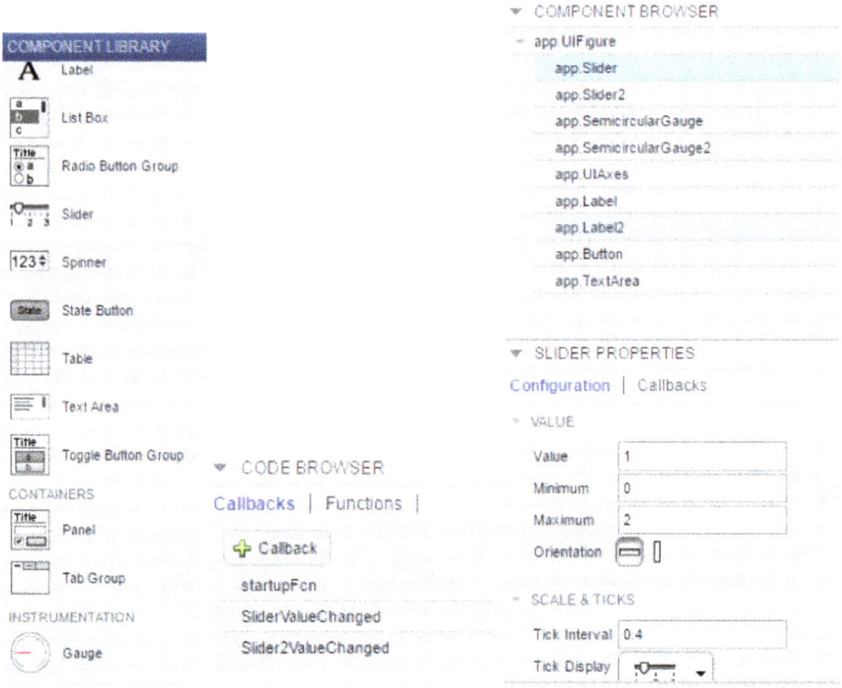

Figure B.2: Some of the app windows. There is a design view toggle to a code view. The component library has buttons, sliders, labels, and gauges. There is a set of component properties. The code browser lets one examine the "callbacks" or action taken when a component property changes, for example.

Code is automatically allocated for actions on callback, such as slider change. Examples can be explored by looking at the script used in the body of the text.

Appendix C

Initial MATLAB Scripts

Introduction

"Editor_ex"

```
%
% comments are in green
%
global xx ; % global (common) variables are blue
%
xx = linspace(0,2 *pi,100); % executables are black
%
var = input('Enter a Variable: '); % text strings are purple
fprintf('Print a Variable %g \n',var); % the %g is the print
    location
yesno = menu('Menu Title?','Yes','No'); % input by popup
    pushbutton
%
for i = 1:20 ; % controls are in blue - indented by loops
    yy(i) = i *sin(i);
end
%
% any syntax errors appear in red, with script explanation on
    hover
%
```

"plot_demo"

```
%
% plot demo
%
close all
```

```
clear all
%
x = linspace(-pi,pi);
y = sin(x);
plot(x,y)
title('Function sin(\theta)')
xlabel('\theta')
ylabel('sin(\theta)')
grid
```

"demo_3d"

```
%
close all
clear all
%
% demo for simple 3-d plots
%
figure
[X,Y] = meshgrid(-8:.5:8);
R = sqrt(X.^2 + Y.^2) + eps;
Z = sin(R)./R;
mesh(Z);
%
figure
[X,Y] = meshgrid(-3:.125:3);
Z = peaks(X,Y);
meshc(Z);
%
figure
[X,Y,Z] = peaks(30);
urfc(X,Y,Z);
axis([-3 3 -3 3 -10 5]);
%
figure
[X,Y] = meshgrid(-2:.2:2);
Z = X.*exp(-X.^2 - Y.^2);
[DX,DY] = gradient(Z,.2,.2);
contour(X,Y,Z)
hold on
quiver(X,Y,DX,DY)
hold off
%
```

"template"

```
%
% Program to compute something
%
close all;                  % clear all plots
clear all;                  % Clear memory
help Template2;                 % Print header
%
% put constants here
%
irun = 1;
%
while irun > 0
    kk = menu('Pick Another Option?','Yes','No');
    if kk == 2
        irun = -1;
        break
    end
    if kk == 1
        %
        param = input('Enter a Parameter): ');
        %
        % do calculations as needed
        %
        x = linspace(0,2 .*pi );
        y = sin(x);
        %
        % make plots
        %
        figure
        plot(x,y)
        title(' sin(x) vs. x')
        xlabel('x')
        ylabel('sin(x)')
        %
    end
%
end
%
```

"ode_demo"

```
%
```

```
% ode example
%
global a
%
a = 1;
%
%tspan = linspace(0,10,200);
tspan = [0 10]; % default is 50 steps
[tt,yy] = ode45(@odedemofun,tspan,1.0);
plot(tt,yy)
xlabel('t')
ylabel('y')
%
function dydt = odedemofun(t,y)
global a
%
dydt = - a .* cos(y)* exp(y);
%
end
```

Classical optics

"Spherical_Mirror"

```
%
% Program to Ray Trace the Focus of a Spherical Mirror and
    Look at
% Aberration
%
clear all;
close all;
help Spherical_Mirror; % Clear memory and print header
%
fprintf(' Focal Point, R/2, of a Spherical Mirror,
    Radius = 1 \n');
%
% unit radius, with incident parallel rays from the left
    (x < 0)
%
irun = 1;
%
while irun > 0
    kk = menu('Pick Another Maximum Incident Ray for the
```

```
Mirror?','Yes','No');
    if kk == 2
       irun = -1;
       break
    end
    if kk == 1
        %
        ymax = input('Enter The Maximum Value of the Incident
Parallel Ray y : ');
        yray = linspace(0,ymax, 10);
        xhit = cos(asin(yray)); % the incident rays
        theta = asin(yray);
        xr(1,:) = [-1 -1 -1 -1 -1 -1 -1 -1 -1 -1];
        xr(2,:) = xhit;
        yr(1,:) = yray(:);
        yr(2,:) = yray(:);
        yr(3,:) = 0.0;
        xr(3,:) = (xhit .*tan(2.0 .*theta) - yray) ./(tan(2.0
.*theta));
        %
        figure
        for i = 1:10
            plot(xr(:,i),yr(:,i),'b-')
            plot(xr(2,i),yr(2,i),'b-',xr(3,i),yr(3,i),'b-')
            hold on
        end
        ym = linspace(0,1); % the mirror
        xm = sqrt(1 - ym .^2);
        plot(xm,ym,'r-')
        title('Spherical Mirror Focal Position for Parallel
Incident Rays')
        xlabel('x')
        ylabel('y')
        axis([-0.0 1.0 0 ymax]);
        %
        hold off
    end
end
%
```

"Parabolic_Mirror"

```
%
% look at paraboic mirror - no speherical abberations
```

```
%
close all
clear all
%
yp = linspace(0,8);
a = 2; % focus at (a,0)
xp = (yp .*yp) ./(4 .*a);
%
% parallel beam in - intersect parabola at x
y = linspace(0,9,10);
x = (y .*y) ./(4 .*a);
%
% angle of incidence, normal = [1 , -y/2a]
plot(xp,yp,'-b')
hold on
for i = 1:length(y)
    plot([x(i),x(10)], [y(i) y(i)],':r') ; % incident
    plot([x(i) x(i)+1] , [y(i) y(i)-y(i) ./(2 .*a)],':b');%normal
    plot([x(i) a] , [y(i) 0],'--g'); % focus
    axis([0 9 0 9])
end
title('Paraboic Mirror, Parallel Rays Incident')
xlabel('x') ; ylabel('y')
hold off
```

Uniform electric and magnetic fields

"Uniform_E_Motion"

```
%
% look at trajectory of a charge in a uniform E field -
analytic
%
clear all; % Clear memory
close all
help Uniform_E_Motion; % Print header
%
syms a t z zp b bc zc
%
% dP/dt = b = qEo, P = bt, dbeta/dt = a (NR) = qEo/m,
dbeta/dt (SR)= a/sqrt(a^2+1)
%
fprintf('Charged Particle in Constant Field Eo \n ')
```

```
%
fprintf('dp/dt = q*Eo, p = (q*Eo)*t, beta = p/E = at/sqrt
((at)^2 + 1), a = q*Eo/m \n ')
%
b = a*t/sqrt((a*t)^2+1);
z = int(a*t/sqrt((a*t)^2+1));
z = z - 1/a; % z(0) = 0
pretty(z)
%
fprintf('Taylor Expansion for z \n ')
%
zp = taylor(z,'Order',5);
pretty(zp)
%
fprintf('Classical Non-relativistic Results \n ')
%
bc = a*t        % NR classical values
zc = (a*t*t)/2
%
tt = linspace(0,5,20);
%
a = 1;
%
t = tt;
bb = eval(b); % dzdt
zz = eval(z); % distance
bbc = eval(bc); % NR classical limits
zzc = eval(zc);
%
figure
for i = 1:length(tt)
    plot(tt(i),bb(i),'-bo',tt(i),bbc(i),'r:*')
    title('\beta = v/c as a Function of t in a Uniform
Electric Field ')
    xlabel('ct')
    ylabel('\beta')
    legend('SR','Classical')
    pause(0.5)
    axis([0 5 0 5])
    hold on
end
```

```
hold off
%
figure
for i = 1:length(tt)
    plot(tt(i),zz(i),'-bo',tt(i),zzc(i),'r*:')
    title('Distance in c*t Units as a function of t')
    xlabel('ct')
    ylabel('z')
    legend('SR','Classical')
    pause(0.5)
    axis([0 5 0 14])
    hold on
end
hold off
%
```

"Uniform_By_Symbolic"

```
%
clear all
close all
%
syms ax(t) az(t) r x(t) z(t)
ode1 = diff(ax(t)) == -az(t)/r;
ode2 = diff(az(t)) == ax(t)/r;
odes = [ode1; ode2];
fprintf('Charged Particle in Constant Field By \n ')
fprintf('Initially moving along z axis, direction cosines\n')
init = [ax(0) == 0; az(0) == 1];
[axs(t),azs(t)] = dsolve(odes,init);
axs
azs
fprintf('Positions x and z \n')
x = int(axs)
z = int(azs)
%
r = 1;
tt = linspace(0,1);
```

```
t = tt;
xx = eval(x);
zz = eval(z);
plot(xx,zz)
title('Motion for curvature radius = 1')
xlabel('x')
ylabel('z')
%
```

"Uniform_B_Motion"

```
%
function[x,p] = Uniform_B_Motion(s, xo, po, Bo)
%
% Take a step of arc length s in a uniform magnetic field of
    magnitude Bo
% along the y axis, charge qe, incident position,xo,yo,
    momentum components
% pox, poy poz - exact solutions
%
ptot = sqrt ( po(1) .^2 + po(2) .^2 + po(3) .^2);
%
% no force along y
%
x(2) = xo(2) + (s .*po(2)) ./ptot;
%
% find radius of curvature (signed), andbendangle, Kg, GeV, m, e
%
a = 33.356; % using q = proton charge
rho = (a .*ptot) ./Bo;
phi = s ./rho;
%
% direction cosines rotate by phi, positions correlated
%
cp = cos(phi);
sp = sin(phi);
p(1) = po(1) .*cp + po(3) .*sp;
p(3) = po(3) .*cp - po(1) .*sp;
p(2) = po(2);
x(1) = xo(1) - (rho .*(p(3) - po(3))) ./ptot;
```

```
x(3) = xo(3) + (rho .*(p(1) - po(1))) ./ptot;
%
```

"Uniform_B_Motion_2"

```
%
% Take a step of arc length s in a uniform magnetic field of
    magnitude Bo
% along the y axis, charge qe, incident position,xo, yo,
    momentum components
% pox, poy poz - exact solutions. Movie made
%
Bo = 10;
ss = 60;
po = [ 1 1 0];
xo = [0 0 0];
ptot = sqrt ( po(1) .^2 + po(2) .^2 + po(3) .^2);
s = linspace(0 ,ss, 50);
for i = 1:length(s)
    %
    % no force along y
    %
    y(i) = xo(2) + (s(i) .*po(2)) ./ptot;
    %
    % find radius of curvature (signed), and bend angle, Kg,
      GeV, m, e
    %
    a = 33.356; % using q = proton charge
    rho = (a .*ptot) ./Bo;
    phi = s(i) ./rho;
    %
    % direction cosines rotate by phi, positions correlated
    %
    cp = cos(phi);
    sp = sin(phi);
    p(1) = po(1) .*cp + po(3) .*sp;
    p(3) = po(3) .*cp - po(1) .*sp;
    p(2) = po(2);
    x(i) = xo(1) - (rho .*(p(3) - po(3))) ./ptot;
    z(i) = xo(3) + (rho .*(p(1) - po(1))) ./ptot;
    %
    plot3(x(i) , y(i), z(i),'-b*')
```

```
    title('Three Dimensional Orbit in a By Field')
    xlabel('x'); ylabel('y'); zlabel('z');
    pause(.3)
    axis([ -4 4 0 60 -8 0])
    hold on
end
%
hold off
%
```

"Solenoid_4"

```
%
% Program to plot B field of a single loop, a pair of current
loops, or 4 loops
% numerically - all points using Biot-Savert
%
clear all;
close all;
help Solenoid_4; % Clear memory and print header
%
fprintf(' B Field for 1, 2 or 4 Current Loops Pair, Radius 1,
Separation = 2d \n');
fprintf(' Current Loops in x, y Plane. Theta is the Angle w.
r.t. the z Axis \n');
fprintf(' B Field for a Current Loop, Radius a, for r > > a,
Biot-Savert \n');
fprintf(' One Loop Dipole Moment ~ pi*I*a^2 \n');
%
irun = 1;
%
while irun > 0
    kk = menu('Pick Another Distance Between Loops','Yes',
'No');
    if kk == 2
        irun = -1;
        break
    end
    if kk == 1
        %
        % Set d
        %
```

```
        d_2 = input('Enter the Distance Between Current
Loops ( ~ 0.5): ');
        ihelm = menu('Number of Loops ','1','2','4');
        d = d_2 ./2.0; % half distance
        xx = linspace(-2,2,20);      % radius in units of a
        zz = linspace(-2,2,20);
        phi=linspace(0,2.0 .*pi);  %šource~integration~-~loop
        Bx = zeros(length(xx),length(zz));
        By = zeros(length(xx),length(zz));
        Bz = zeros(length(xx),length(zz));
        %
        for i = 1:length(xx)
            for j = 1:length(zz)        % grid of field points
                for k = 1:length(phi)  % integrate over source
                    cp = cos(phi(k));
                    sp = sin(phi(k));
                    rr32p = (xx(i) .^2 + 1 - 2.0 .*xx(i)
                    .*cp + (zz(j)-d) .^2) .^1.5 ;
                    rr32p2 = (xx(i) .^2 + 1 - 2.0 .*xx(i)
                    .*cp + (zz(j)- 3 .*d) .^2) .^1.5 ;
                    rr32m = (xx(i) .^2 + 1 - 2.0 .*xx(i)
                    .*cp + (zz(j)+d) .^2) .^1.5 ;
                    rr32m2 = (xx(i) .^2 + 1 - 2.0 .*xx(i)
                    .*cp + (zz(j)+ 3 .*d) .^2) .^1.5 ;
                    dBxp = cp .*(zz(j) - d) + d .*sp;
                    dBxm = cp .*(zz(j) + d) - d .*sp;
                    dBxp2 = cp .*(zz(j) - 3 .*d) + d .*sp;
                    dBxm2 = cp .*(zz(j) + 3 .*d) - d .*sp;
                    dByp = d .*(xx(i) - cp) + (zz(j) - d)
                    .*sp;
                    dBym = -d .*(xx(i) - cp) + (zz(j) + d)
                    .*sp;
                    dByp2 = d .*(xx(i) - cp) + (zz(j) - 3 .*d)
                    .*sp;
                    dBym2 = - d .*(xx(i) - cp) + (zz(j) + 3
                    .*d) .*sp;
                    dBzp = (1.0 - xx(i) .*cp);
                    dBzm = dBzp;
                    % single loop at z = d
                    if ihelm == 1 ; % single loop
                        Bx(i,j) = Bx(i,j) + (dBxp ./rr32p);
                        By(i,j) = By(i,j) + (dByp ./rr32p);
```

```
                Bz(i,j) = Bz(i,j) + (dBzp ./rr32p);
            end
            % current sense is the same for 4 loops -
            reinforcing
            if ihelm == 2 ; % 2 loops Helmholtz
                Bx(i,j) = Bx(i,j) + (dBxp ./rr32p +
                dBxm ./rr32m);
                By(i,j) = By(i,j) + (dByp ./rr32p +
                dBym ./rr32m);
                Bz(i,j) = Bz(i,j) + (dBzp ./rr32p +
                dBzm ./rr32m);
            end
            if ihelm == 3 ; % 4 loops, ~ solenoid
                Bx(i,j) = Bx(i,j) + (dBxp ./rr32p +
                dBxm ./rr32m);
                By(i,j) = By(i,j) + (dByp ./rr32p +
                dBym ./rr32m);
                Bz(i,j) = Bz(i,j) + (dBzp ./rr32p +
                dBzm ./rr32m);
                Bx(i,j) = Bx(i,j) + (dBxp2 ./rr32p2 +
                dBxm2 ./rr32m2);
                By(i,j) = By(i,j) + (dByp2 ./rr32p2 +
                dBym2 ./rr32m2);
                Bz(i,j) = Bz(i,j) + (dBzp ./rr32p2 +
                dBzm ./rr32m2);
            end
        end
    end
end
%
figure
contour(xx,zz,Bz',40);
xlabel('x/a')
ylabel('z/a')
title('Contour for Bz')
%
figure
mesh(xx,zz,Bz');
xlabel('x/a')
ylabel('z/a')
title('Mesh for Bz')
%
```

```
        figure
        contour(xx,zz,Bx',40);
        xlabel('x/a')
        ylabel('z/a')
        title('Contour for Bx')
    end
end
%
```

"Cyclotron_NR"

```
%
% Program to look at Cyclotron Operation - NR only - p or ion
%
clear all;
close all;
help Cyclotron_NR     % Clear the memory, figures and print
header
%
% work in dimensionless units as possible
%
fprintf('Cyclotron w = qB/m, r = m*v/qB = v/w, E = (qBr)
^2/2m \n')
fprintf('Relativistic Effects, m -> E, w Decreases and r
Increases by gamma \n')
%
irun = 1;
%
while irun > 0
    kk = menu('Pick Another Number of Half Rotations',
'Yes','No');
    if kk == 2
        irun = -1;
        break
    end
    if kk == 1
        %
        n = input('Enter Number of Half Rotations, ~ 40: ');
        dE = input('Enter Energy Kick in Cross Dees ( ~ 0.5,
           dE~ rdr): ');
        %
        wt = linspace(0.0, pi .*n,200);
```

```
% initial radius
a(1) = 0.05;    % initial energy/radius - p source
%
xx(1) = a(1);   % starting up
yy(1) = 0;
dwt = wt(2)- wt(1);
imax = length(wt);
for i = 2:imax
    j = round(1 +(i *n ./imax));    % which half
      rotation?
    E(j) = j .*dE;   % kinetic energy gains dE for
      each i/2 rotation
    a(j) = sqrt(E(j)); % dE ~ rdr, radius for this
      1/2 rotation
    xx(i) = xx(i-1) + a(j) .*cos(wt(i)) .*dwt;
    yy(i) = yy(i-1) + a(j) .*sin(wt(i)) .*dwt;
    if sqrt(xx(i) .^2 + yy(i) .^2) > 5 ;   % set
       radius of the Dees
        fprintf('Exit Dees \n')
        imax = i;
        break
    end
end
%
% draw "dees"
%
xdee = linspace(-5,5);
ydee = sqrt(5.0 .^2 - xdee .^2);
yp = [0.1 0.1]; ym = [- 0.1 -0.1];
xp = [-5.0 5.0]; xm = [-5.0 5.0];
figure;
for i = 1:imax
    plot(xdee,ydee,'b',xdee,-ydee,'r',xp,yp,'b-',
      xm,ym,'r-',xx(i),yy(i),'-og')
    xlabel('x(m)')
    ylabel('y(m)')
    title('Cyclotron Orbits - Non-Relatisivtic')
    axis([ -5 5 -5 5])
    pause(0.05)
end
plot(xdee,ydee,'b',xdee,-ydee,'r',xp,yp,'b-',xm,ym,
  'r-',xx,yy,'og')
```

```
        xlabel('x(m)')
        ylabel('y(m)')
        title('Cyclotron Orbits - Non-Relatisivtic')
        axis([ -5 5 -5 5])
    end
end
%
```

Appendix D

Useful Trig Identities

There are several occasions to use trig identities in the text, especially in the sections on periodic assemblies. A table is provided here in order to make more accessible the calculations performed in those cases. The MATLAB symbolic math package does not always apply these identities optimally. The user may need to use the "subs" utility to impose the identity explicitly.

Pythagorean Identities

$$\sin^2\theta + \cos^2\theta = 1 \qquad \sec^2\theta = 1 + \tan^2\theta \qquad \csc^2\theta = 1 + \cot^2\theta$$

Quotient Identities

$$\tan\theta = \frac{\sin\theta}{\cos\theta} \qquad \cot\theta = \frac{\cos\theta}{\sin\theta}$$

Sum or Difference of Two Angles

$$\sin(\alpha \pm \beta) = \sin\alpha\cos\beta \pm \cos\alpha\sin\beta$$
$$\cos(\alpha \pm \beta) = \cos\alpha\cos\beta \mp \sin\alpha\sin\beta$$
$$\tan(\alpha \pm \beta) = \frac{\tan\alpha \pm \tan\beta}{1 \mp \tan\alpha\tan\beta}$$

Product to Sum Formulas

$$\cos\alpha\cos\beta = \tfrac{1}{2}(\cos(\alpha-\beta)+\cos(\alpha+\beta))$$
$$\sin\alpha\sin\beta = \tfrac{1}{2}(\cos(\alpha-\beta)-\cos(\alpha+\beta))$$
$$\sin\alpha\cos\beta = \tfrac{1}{2}(\sin(\alpha+\beta)+\sin(\alpha-\beta))$$
$$\cos\alpha\sin\beta = \tfrac{1}{2}(\sin(\alpha+\beta)-\sin(\alpha-\beta))$$

Double Angle Formulas

$$\sin 2\theta = 2\sin\theta\cos\theta \qquad \tan 2\theta = \frac{2\tan\theta}{1-\tan^2\theta}$$
$$\cos 2\theta = \cos^2\theta - \sin^2\theta$$
$$\cos 2\theta = 2\cos^2\theta - 1$$
$$\cos 2\theta = 1 - 2\sin^2\theta$$

Sum to Product Formulas

$$\sin\alpha \pm \sin\beta = 2\sin\left(\frac{\alpha\pm\beta}{2}\right)\cos\left(\frac{\alpha\mp\beta}{2}\right)$$
$$\cos\alpha + \cos\beta = 2\cos\left(\frac{\alpha+\beta}{2}\right)\cos\left(\frac{\alpha-\beta}{2}\right)$$
$$\cos\alpha - \cos\beta = -2\sin\left(\frac{\alpha+\beta}{2}\right)\sin\left(\frac{\alpha-\beta}{2}\right)$$

Half-Angle Formulas

$$\sin\frac{\theta}{2} = \pm\sqrt{\frac{1-\cos\theta}{2}} \qquad \cos\frac{\theta}{2} = \pm\sqrt{\frac{1+\cos\theta}{2}}$$
$$\tan\frac{\theta}{2} = \csc\theta - cot\theta \qquad \cot\frac{\theta}{2} = \csc\theta + cot\theta$$
$$\tan\frac{\theta}{2} = \pm\sqrt{\frac{1-\cos\theta}{1+\cos\theta}} \qquad \tan\frac{\theta}{2} = \pm\sqrt{\frac{1+\cos\theta}{1-\cos\theta}}$$
$$\tan\frac{\theta}{2} = \frac{\sin\theta}{1+\cos\theta} \qquad \tan\frac{\theta}{2} = \frac{\sin\theta}{1-\cos\theta}$$
$$\tan\frac{\theta}{2} = \frac{1-\cos\theta}{\sin\theta} \qquad \tan\frac{\theta}{2} = \frac{1+\cos\theta}{\sin\theta}$$

Co-Function Identities

$$\sin\left(\tfrac{\pi}{2}-\theta\right) = \cos\theta \qquad \csc\left(\tfrac{\pi}{2}-\theta\right) = \sec\theta$$
$$\cos\left(\tfrac{\pi}{2}-\theta\right) = \sin\theta \qquad \sec\left(\tfrac{\pi}{2}-\theta\right) = \csc\theta$$
$$\tan\left(\tfrac{\pi}{2}-\theta\right) = \cot\theta \qquad \cot\left(\tfrac{\pi}{2}-\theta\right) = \tan\theta$$

Even-Odd Identities

$$\sin(-\theta) = -\sin\theta \qquad \csc(-\theta) = -\csc\theta$$
$$\cos(-\theta) = \cos\theta \qquad \sec(-\theta) = \sec\theta$$
$$\tan(-\theta) = -\tan\theta \qquad \cot(-\theta) = -\cot\theta$$

Figure D.1: Trigonometric identities of use in the algebraic derivations.

Appendix E

Collider Accelerator Parameters

The table referring to Fermilab accelerators is taken from Ref. 2. It refers to the Tevatron in collider mode when CDF and D0 are taking data as a 1.8 TeV CM collider. Other tables in Ref. 2 give Booster parameters and 150 GeV Main Ring parameters in operation as injector into the Tevatron. Parameters for the Fermilab Linac are also shown.

Table A.3. Tevatron — collider mode.

Circumference	$2\pi \times 1000$ meters
Injection energy	150 GeV
Peak energy	900 GeV
Acceleration period	52 sec
Harmonic number, h	1113
Transition gamma	18.7
Maximum RF voltage	1.4 MV
Longitudinal emittance	3 eV sec
β_{max} in insertion	900 meters
β_{max} in cells	100 meters
β^* at collision point	0.5 meter
Maximum dispersion	12 meters
Tune $\nu_x \approx \nu_y$	19.4
Transverse emittance*	24π mm mrad
Bend magnet length	6.1 meters
Standard half-cell length	29.7 meters
Bend magnets per cell	8
Bend magnet total	774
Typical bunch intensity:	
protons	1×10^{11}
antiprotons	5×10^{10}
Phase advance per cell	68 deg
Cell type	FODO

Fermilab Linac

Beam Energy Range	100-400 MeV
Intensity Control	few protons/pulse to 4.2×10^{10} protons/pulse
Emittance Selection	0.1π to 6π mm-mrad
Pulse Length	picosecs to 40 µsec
Transverse Beam Size	.5 mm to 75 mm
Momentum Spread, $\Delta p/p$	0.3% to .05%

Figure E.1: Tabulated data for the Tevatron in collider mode and the Fermilab Linac.

Fermilab Booster	
Circumference (m)	474.2
Average machine radius (m)	75.47
Injection kinetic energy (MeV)	400
Extraction kinetic energy (GeV)	8
RF frequency (MHz)	37.87/52.8
Harmonic number	84
Protons per bunch	6 × 10^10
Max/Min βx (m)	33.67/20.46
Phase advance per cell ϕ (degree)	102
Horizontal, vertical tune ν , ν	6.7 6.8
x y Transition γ	5.45
Revolution frequency at injection, extraction (kHz)	451, 629
Normalized transverse emittance εN (95%, mm-mrad)	12 π
Longitudinal emittance (95%, eV-s)	0.1

Figure E.2: Parameters of the Fermilab Booster.

LHC parameters related to the bunch structure of the beams and LHC luminosity. The Super Proton Synchrotron (SPS) is the injector for the LHC.

		Injection	Collision
Beam Data			
Proton energy	[GeV]	450	7000
Relativistic gamma		479.6	7461
Number of particles per bunch		1.15×10^{11}	
Number of bunches		2808	
Longitudinal emittance (4σ)	[eVs]	1.0	2.5[a]
Transverse normalized emittance	[μm rad]	3.5[b]	3.75
Circulating beam current	[A]	0.584	
Stored energy per beam	[MJ]	23.3	362
Peak Luminosity Related Data			
RMS bunch length[c]	cm	11.24	7.55
RMS beam size at the IP1 and IP5[d]	μm	375.2	16.7
RMS beam size at the IP2 and IP8[e]	μm	279.6	70.9
Geometric luminosity reduction factor F[f]		-	0.836
Peak luminosity in IP1 and IP5	[$\mathrm{cm^{-2}sec^{-1}}$]	-	1.0×10^{34}
Peak luminosity per bunch crossing in IP1 and IP5	[$\mathrm{cm^{-2}sec^{-1}}$]	-	3.56×10^{30}

Figure E.3: Tabulated data for the LHC at injection energy and at maximum collider energy.

LHC parameters related to the geometry, lattice, and RF acceleration systems.

		Injection	Collision
Geometry			
Ring circumference	[m]	26658.883	
Ring separation in arcs	[mm]	194	
Bare inner vacuum screen height in arcs	[mm]	46.5	
Effective vacuum screen height (incl. tol.)	[mm]	44.04	
Bare inner vacuum screen width in arcs	[mm]	36.9	
Effective vacuum screen width (incl. tol.)	[mm]	34.28	
Main Magnet			
Number of main bends		1232	
Length of main bends	[m]	14.3	
Field of main bends	[T]	0.535	8.33
Bending radius	[m]	2803.95	
Lattice			
Maximum dispersion in arc	[m]	2.018 (h) / 0.0 (v)	
Minimum horizontal dispersion in arc	[m]	0.951	
Maximum β in arc	[m]	177 (h)/ 180 (v)	
Minimum β in arc	[m]	30 (h) / 30 (v)	
Horizontal tune		64.28	64.31
Vertical tune		59.31	59.32
Momentum compaction	10^{-4}	3.225	
Slip factor η	10^{-4}	3.182	3.225
Gamma transition γ_{tr}		55.68	
RF System			
Revolution frequency	[kHz]	11.245	
RF frequency [a]	[MHz]	400.8	
Harmonic number		35640	
Number of bunches		2808	
Total RF voltage	[MV]	8	16
Synchrotron frequency	[Hz]	61.8	21.4
Bucket area	[eVs]	1.46	8.7
Bucket half height ($\Delta E/E$)	[10^{-3}]	1	0.36

Figure E.4: LHC parameters for the dipoles, FODO unit cell, and RF acceleration system.

	HERA (DESY)	TEVATRON* (Fermilab)	RHIC (Brookhaven)	LHC (CERN)		
Physics start date	1992	1987	2001	2009	2015	2026 (HL-LHC)
Physics end date	2007	2011	—	—		
Particles collided	ep	$p\bar{p}$	pp (polarized)	pp		
Maximum beam energy (TeV)	e: 0.030 p: 0.92	0.980	0.255 53% polarization	4.0	6.5	7.0
Maximum delivered integrated luminosity per exp. (fb^{-1})	0.8	12	0.38 at 100 GeV 0.75 at 250/255 GeV	23.3 at 4.0 TeV 6.1 at 3.5 TeV	4.2	250/y
Luminosity (10^{30} $cm^{-2}s^{-1}$)	75	431	245 (pk) 160 (avg)	7.7×10^3	5×10^3	5.0×10^4 (leveled)
Time between collisions (ns)	96	396	107	49.90	24.95	24.95
Full crossing angle (μ rad)	0	0	0	290	290	590
Energy spread (units 10^{-3})	e: 0.91 p: 0.2	0.14	0.15	0.1445	0.105	0.123
Bunch length (cm)	e: 0.83 p: 8.5	p: 50 $\bar{p}$: 45	60	9.4	9	9
Beam radius (10^{-6} m)	e: 110(H),30(V) p: 111(H),30(V)	p: 28 $\bar{p}$: 16	85	18.8	21	7
Free space at interaction point (m)	±2	±6.5	16	38	38	38
Initial luminosity decay time, $-L/(dL/dt)$ (hr)	10	6 (avg)	7.5	≈ 6	≈ 20	≈ 6 (leveled)
Turn-around time (min)	e: 75, p: 135	90	25	180	134	180
Injection energy (TeV)	e: 0.012 p: 0.040	0.15	0.023	0.450	0.450	0.450
Transverse emittance (10^{-9} m)	e: 20(H),3.5(V) p: 5(H),5(V)	p: 3 $\bar{p}$: 1	13	0.59	0.5	0.34
β^*, ampl. function at interaction point (m)	e: 0.6(H),0.26(V) p: 2.45(H),0.18(V)	0.28	0.65	0.6	0.8	0.15
Beam-beam tune shift per crossing (units 10^{-4})	e: 190(H),450(V) p: 12(H),9(V)	p: 120 $\bar{p}$: 120	73	72	37	110
RF frequency (MHz)	e: 499.7 p: 208.2/52.05	53	accel: 9 store: 28	400.8	400.8	400.8
Particles per bunch (units 10^{10})	e: 3 p: 7	p: 26 $\bar{p}$: 9	18.5	16	12	22
Bunches per ring per species	e: 189 p: 180	36	111	1380	2244 2232 (i.r. 1/5†)	2748 2736 (i.r. 1/5†)
Average beam current per species (mA)	e: 40 p: 90	p: 70 $\bar{p}$: 24	257	400	467	1200
Circumference (km)	6.336	6.28	3.834	26.659		
Interaction regions	2 colliding beams 1 fixed target (e beam)	2 high $\mathcal{L}$	6 total, 2 high $\mathcal{L}$	4 total, 2 high $\mathcal{L}$		
Magnetic length of dipole (m)	e: 9.185 p: 8.82	6.12	9.45	14.3		
Length of standard cell (m)	e: 23.5 p: 47	59.5	29.7	106.90		
Phase advance per cell (deg)	e: 60 p: 90	67.8	84	90		
Dipoles in ring	e: 396 p: 416	774	192 per ring + 12 common	1232 main dipoles		
Quadrupoles in ring	e: 580 p: 280	216	246 per ring	482 2-in-1 24 1-in-1		
Magnet types	e: C-shaped p: s.c., collared, warm iron	s.c., cos θ warm iron	s.c., cos θ cold iron	s.c., 2 in 1 cold iron		
Peak magnetic field (T)	e: 0.274, p: 5	4.4	3.5	8.3‡		

Figure E.5: Summary table of hadron collider parameters from the Particle Data Group.

C. Patrignani *et al.* (Particle Data Group), *Chin. Phys. C*, **40**, 100001 (2016). http://pdg.lbl.gov/2016/reviews/contents_ sports.html

Appendix F

Radiation Length of the Elements

Element	Z	A [g/mol]	Rad. Length (expt.) [g.cm^{-2}]	Rad. Length (analyt.) [g.cm^{-2}]	Error [%]
H	1	1.00794	63.04	63.79	1.2
He	2	4.0026	94.32	89.95	4.63
C	6	12.0108	42.7	43.01	0.72
N	7	14.0067	37.99	38.23	0.64
O	8	15.9994	34.24	34.46	0.64
F	9	18.9984	32.93	33.16	0.69
Ne	10	20.1797	28.93	29.15	0.77
Na	11	22.9897	27.74	27.97	0.84
Mg	12	24.305	25.03	25.27	0.96
Al	13	26.9815	24.01	24.26	1.06
Si	14	28.0855	21.82	22.08	1.18
P	15	30.9737	21.21	21.47	1.25
S	16	32.065	19.5	19.76	1.35
Cl	17	35.453	19.28	19.56	1.47
Ar	18	39.948	19.55	19.86	1.57
K	19	39.0983	17.32	17.6	1.64
Ca	20	40.078	16.14	16.43	1.78
Ti	22	47.867	16.16	16.47	1.94
Cr	24	51.9961	14.94	15.25	2.09
Fe	26	55.845	13.84	14.14	2.17
Ni	28	58.6934	12.68	12.97	2.27
Cu	29	63.546	12.86	13.16	2.34
Zn	30	65.38	12.43	12.72	2.35
Ag	47	107.868	8.97	9.17	2.26
Pt	78	195.084	6.54	6.52	0.38
Au	79	196.967	6.46	6.43	0.53
Pb	82	207.2	6.37	6.31	0.93

Figure F.1: Radiation length of the elements.

References

[1] E.D. Courant and H.S. Snyder, *Phys. Rev.* **88**, 1190 (1952); Theory of the Alternating-Gradient Synchrotron, *Ann. Phys.* **281**, 360–408 (2000).
[2] D.A. Edwards and M.J. Syphers, *An Introduction to the Physics of High Energy Accelerators* (Wiley, 1993). Published online: 23 Jan. 2008.
[3] S.Y. Lee, *Accelerator Physics* (World Scientific, 1999).
[4] E.J.N. Wilson, *An Introduction to Particle Accelerators* (Oxford University Press, 2001).
[5] M. Conte and W. MacKay, *Introduction to the Physics of Particle Accelerators* (World Scientific, 1991).
[6] D.C. Carey, *The Optics of Charged Particle Beams* (Harwood Academic, 1987).
[7] Particle Accelerator schools: http://uspas.fnal.gov and http://cas.web.cern.ch.
[8] A.W. Chao and M. Tigner, *Handbook of Accelerator Physics and Engineering*, 3rd printing (World Scientific, Singapore, 2006).
[9] References from the US Particle Accelerator School (USPAS) site:

Accelerator Physics	S.Y. Lee	**Accelerator Physics — third edition** (World Scientific, 2012)
Accelerator Physics	Edmund Wilson	**An Introduction to Particle Accelerators** (Oxford University Press, 2001)
Accelerator Physics	Donald A. Edwards and Michael J. Syphers	**An Introduction to the Physics of High Energy Accelerators** (John Wiley & Sons, 1992)
Accelerator Physics	Mario Conte and William M. MacKay	**An Introduction to the Physics of Particle Accelerators — second edition** (World Scientific, 2008)
Accelerator Physics	Ken Takayama and Richard J. Briggs (eds.)	**Induction Accelerators** (Springer, 2010)
Accelerator Physics	Helmut Wiedemann	**Particle Accelerator Physics — fourth edition** (Springer, 2015)
Accelerator Physics	Stanley Humphries, Jr.	**Principles of Charged Particle Acceleration** (Wiley, 1986)
Accelerator Physics	John Jacob Livingood	**Principles of Cyclic Particle Accelerators** (Van Nostrand, NJ, 1961)
Accelerator Physics	Thomas P. Wangler	**RF Linear Accelerators — second, completely revised and enlarged edition** (Wiley, 2008)

Index

www.ingramcontent.com/pod-product-compliance
Lightning Source LLC
Chambersburg PA
CBHW050556190326
41458CB00007B/2066
9789813237469